Fire on the Land

A Retrospective Anthology of
Selected Papers from the Archives
of the Society of American Foresters

The Society of American Foresters

forests. resources. communities.

The Society of American Foresters
10100 Laureate Way
Bethesda, Maryland 20814
United States
www.eforester.org

Disclaimer: This book and contents included within do not constitute professional advice and should not be taken as such.

Printed in the United States of America on paper certified by the Sustainable Forestry Initiative. The paper used in this printing meet the requirements of ANSI/NISO Z39.48-1992 (Permanence of Paper).

ISBNs: 978-0-939970-32-2 (paperback)
        978-0-939970-34-6 (eBook)

Cover photos (front and back) by Stephen Fillmore, USDA Forest Service, 2015.

About the Society of American Foresters

The capability of foresters and natural resource professionals to continue to provide forest benefits to the public is vital to the environment, social, and economic health of the United States and the broader global community.

Founded in 1900 by Gifford Pinchot, the Society of American Foresters (SAF) is the national scientific and educational organization representing the forestry profession in the United States. It is the largest professional society for foresters in the world. The mission of SAF is to advance sustainable management of forest resources through science, education, and technology; to enhance the competency of its members; to establish professional excellence; and to use our knowledge, skills, and conservation ethic to ensure the continued health, integrity, and use of forests to benefit society in perpetuity. SAF is a nonprofit organization meeting the requirements of 501(c)(3). SAF members include natural resource professionals in public and private settings, researchers, CEOs, administrators, educators, and students.

A Note About the Cover Art

The photograph I chose for the cover of *Fire on the Land* is a picture I took while out on the Lake Fire in 2015. I call it "Tree and Cone." I've loved this picture ever since I was lucky enough to capture it. It hints at past management (note the old saw cut top-left) as well as the old burn marks distributed around this piece of wood (an old limber pine relic). The cone nestled into the dead wood signifies the birth–death cycle and all in the context of wildland fire management. Simple…

The image used on the back cover is from the same fire. I like the movement of fire in this picture, swirling around the bole of the tree. I also like the intact crown structure of the tree; it is unclear whether the tree will fight off the flames or succumb to them.

—Stephen Fillmore, Editor

Contents

Part 5—The Utilization of Fire 203

Introduction by Kelly Martin

Pivotal Papers from the SAF Archive

List of Contributors

Editor
Stephen D. Fillmore
Cleveland National Forest
United States Forest Service
San Diego, California

Contributors
Dr. Ross W. Gorte
Retired, Natural Resources Section,
Congressional Research Service
Library of Congress
Arvada, Colorado

Kelly Martin
National Park Service
Yosemite, California

Stephen J. Pyne
Professor, School of Life Science
Arizona State University
Tempe, Arizona

Dr. Elizabeth Reinhardt
Retired, United States Forest Service
Arlee, Montana

Dr. Alistair M.S. Smith
Professor of Wildland Fire Science and
Combustion Physics
College of Natural Resources
University of Idaho
Moscow, Idaho

Brian Van Winkle
Dixie and Fishlake National Forests
United States Forest Service
Cedar City, Utah

Managing Editor
Jennifer Kuhn
Society of American Foresters
Bethesda, Maryland

About the Contributors

Stephen D. Fillmore has been directly involved in fire and fuels management for almost 20 years, having received his start with his universities' volunteer wildland fire crews. Seeing that fuels management was the perfect combination of forestry and wildfire, he has spent his entire professional life directly involved in it, serving in three different fuels positions in both California and Idaho. He is active in wildland fire training including leading a regional cadre for RX-310. He is also active in wildland fire management, holds incident qualifications including RXB2, ICT3, DIVS, FOBS, and FEMO, and serves on a federal Incident Management Team.

Ross W. Gorte has a B.Sc. in forest management from Northern Arizona University, 1975; an M.B.A. from Northern Arizona University, 1976; and a Ph.D. in natural resource economics from Michigan State University, 1981. Ross is a retired senior policy specialist analyst in the Natural Resources Section, Congressional Research Service, Library of Congress; he is currently an affiliate research professor with the Earth Systems Research Center, University of New Hampshire. CRS provides objective, nonpartisan information and analysis on federal policies and programs and on proposed changes for the members and committees of congress. Ross covered forestry and federal land management, including wilderness and other special designations; timber industry, taxation, and trade; wildfire control and effects; and federal appropriations and budgets for and economic impacts of federal land management. He has authored or coauthored more than 100 articles and reports for congress, and has testified on these issues before several congressional committees.

Kelly Martin is the chief of fire and aviation management for the National Park Service at Yosemite National Park. She has been managing and leading the wildland fire program for the last 12 years. Prior to working for the NPS, she spent the prior 16 years working in wildland fire management for the USFS throughout the intermountain west as a fire management officer on 3 different National Forests. She has a bachelor's degree from Northland College and has completed technical fire management. She currently chairs the national wildfire coordinating group for national fire management, a training course designed to engage federal agency administrators from throughout the United States in risk and decision making in complex fire management programs.

Elizabeth Reinhardt is recently retired after 34 years with the U.S. Forest Service. She spent much of her career at the Missoula fire lab, where she conducted research in fuels and fire effects. At retirement she was the assistant director of fire and aviation management for the U.S. Forest Service, with national responsibility for fire ecology and fuel management. She has degrees in English (Harvard, B.A.) and forestry (University of Montana, M.S., Ph.D). She lives near Arlee, Montana.

Alistair Smith is a professor of wildland fire science and combustion physics at the University of Idaho. He received his B.Sc. in physics from the University of Edinburgh, a M.Sc. in physics from the University of London, and a Ph.D in geography from the University of London. Professor Smith has published more than 100 scholarly works focused on wildland fire science, fuels characterization, and fire emissions.

Stephen Pyne spent 15 seasons with the North Rim Longshots at Grand Canyon National Park, then wrote fire plans for the National Park Service for three years. He has been a professor at Arizona State University since 1985. He has written over 30 books, most on the history and management of fire, including historical surveys of the U.S., Australia, Canada, Europe, Russia, and the Earth overall. His most recent fire works include *Between Two Fires: A Fire History of Contemporary America,* and *To the Last Smoke: A Suite of Regional Fire Surveys.*

Brian Van Winkle has worked for the U.S. Forest Service for fourteen years on a variety of national forests in Illinois, Utah, Arkansas, Arizona, and Idaho where he has held positions in trails, fire suppression, timber management, and fuels management. He is currently a fire ecologist on multiple national forests in Utah. He holds a B.Sc. in forestry from Southern Illinois University.

Preface

Stephen Pyne

American Forestry, American Fire

> "American foresters have found that they have a unique fire problem, and they can get little help in solving it from European foresters..."

> – Coert duBois (1914)

By the late 19th century America's fire scene resembled Brazil's a century later. It was aglow with routine folk burning, slash-powered wildfires, and the occasional conflagration. Bernhard Fernow dismissed the spectacle as one of "bad habits and loose morals." Many academic foresters did not even consider fire protection as a part of their craft: fire protection was a precondition for forestry, not one of its endless chores.

But Americans could not avoid the issue. In 1910 Henry Graves, then chief forester, proclaimed that fire protection was 90% of American forestry. As Coert duBois put it, "We must work it out for ourselves. We must gather our own facts, arrange them according to our own classifications, and draw our own conclusions." To a remarkable degree, they did. The response, what came to be called systematic fire protection, would be one of America's canonical contributions to forestry and to forest conservation.

Firefighting became the romance of American forestry. In many ways, a national firefight made a bolder, cleaner, and more coherent narrative of forestry's founding than the complexities of taxes, land laws, and politics during the Gilded Age (and later the Progressive Era). The challenge of state-sponsored conservation could be simplified into a story of an easily understood struggle, if not between good and evil, then at least between abusive burning and those who sought to halt its depredations.

Like all creation stories, this one has a basis in truth, and it deserves a wider context to prevent its descent into anecdote and myth.

***

State-sponsored conservation, with foresters as its designated agents, was a global project. All the imperial powers encouraged it, but perhaps none so robustly as Great Britain. Three German foresters pioneered the program: Dietrich Brandis, Berthold Ribbentrop, and William Schlich (of the *Manual of Forestry* and editor of the *Indian*

Forester). Together they transplanted a species of European agronomy evolved in central Europe, a region without a natural basis for fire, into a host of fire-flushed landscapes.

The preferred technique was the forest reservation, which could then be spared from ax, hoof, and fire. Techniques included preseason preparations like "early burning" of fuelbreaks, followed by rapid detection and attack. The model was pursued in bush, veldt, jungle, and wilderness throughout the empire and had wide appeal. When Rudyard Kipling wrote a sequel to the *Jungle Book*, he had Mowgli, now grown, join the Indian Forest Department as a fire guard. It's worth recalling that both Gifford Pinchot and Henry Graves studied under Brandis, with Graves including a tour of British India.

But fire proved formidable. The first forestry conference in British India, in 1878, opened with two questions upon which everything else depended. Is fire control possible? and, if possible, is it desirable? The responses differed. Academics, administrators, and progressive thinkers all insisted that fire control was both feasible and essential. Open fire was a stigma of primitivism; fire control, the badge of modernity. Field practitioners and guards argued that fire protection on the European model was an illusion and one that would damage the woods and sour relations with the natives. They are questions and, in some ways, answers that remain today.

When Americans took up the task, they recognized their circumstances were different, but they were determined to tame landscape fire. Gifford Pinchot early picked fire as a public symbol of what threatened America's forests. It was graphic, easily understood, and made for perfect publicity. Then, after the Great Fires of 1910 engulfed the fledgling USDA Forest Service, the agency doubled down on fire. Coert duBois codified systematic fire protection,

foresters condemned light burning, and fire protection became the most visible emblem of rational forest management.

Curiously, Americans were strengthening their push ahead just as British India's foresters were pulling back. After decades of expensive efforts with teak, sal, and chir pine, it was obvious that European methods had failed, that indigenous practices—lubricated with abundant burning—had succeeded. Reluctantly, British foresters began compromising with fire, which they never liked but had to accept. In 1905, Burma withdrew formal fire protection altogether.

By then, American foresters were consumed with their own projects and were not inclined to compromise. During the 1930s, American forestry reached the height of its political power, and with the might of the Civilian Conservation Corps it intensified its firefight. The new marker was the 10 a.m. policy announced in 1935, an "experiment on a continental scale," as Chief Forester Gus Silcox put it, that would break the back of widespread burning. A world war and the Cold War reinforced the sentiment that the great firefight was a species of national defense.

But the US, too, faced a fire insurgency, with Herbert Stoddard, Ed Komarek, Harold Weaver, and Harold Biswell as its most public faces. In 1962, a fire revolution that sought to restore good fire trumpeted its opening fanfares with the first Tall Timbers Fire Ecology Conference and the Nature Conservancy's inaugural prescribed fire. Burning for wildlife, burning for ecological services and biodiversity, burning to reinstate natural processes in wilderness, not just hazard reduction—these were the new goals. By 1968, the National Park Service reformed its fire policy accordingly and, in 1978, the Forest Service followed.

Too often forestry found itself on the wrong side of history and was pushed to the margins. Americans were not alone:

imperial forestry and its institutions receded with the decay of empires everywhere in the postwar era. Meanwhile, an Anthropocene was putting fire, both landscape burning and industrial combustion, at the core of global change such that every aspect of forests and public lands generally seemed to be touched by it. An impartial observer might wonder if, after a century, fire is once again 90% of American public forestry.

After 1910, we spent 50 years trying to take out all fire and then spent 50 years trying to put back good fire. It's been an oft-wrenching experience. We now seem to have evolved into a new phase in which the treatment of choice is a hybrid of suppression and prescribed burning—a managed wildfire. We'll see.

What can we learn from this history? How difficult it is to invent a new idea and put it in the field. How hard it is to admit you're wrong. How critical it is that reforms come from within the ranks rather than be imposed from outside. How much boldness must meld with humility, and insight with practice. and, if the past is prologue, as Shakespeare held, how much wildland fire management will likely remain one of American forestry's great experiments.

Introduction

Stephen Fillmore

On behalf of the contributing authors in this book, both past and present, and the Society of American Foresters (SAF), I am pleased to present this collection of important papers from the archives of SAF's journals relating to fire in the natural environment. No organization has a richer or more relevant publishing history than SAF when tasked with a critical interpretation of past events in wildland fire and forestry in the United States and what that means in the context of modern and future practice. As stewards of the future of one of our nation's greatest natural resource, our forests, we must seek to fully understand the past to glean better insights into how to manage the challenges that lie ahead. This anthology seeks to bridge the discussions of bygone decades with implications for today, and an understanding of what changes may be warranted in the years and decades yet to be.

The venerable *Journal of Forestry* celebrates over 115 years of continuous publication. This legacy, rich with history, is an honor that not many organizations or publications in the field of forestry can claim. As such, SAF and its publications have borne witness to policies that have come and gone, scientists who have entered the fray and retired, fires lit and fires extinguished.

Of course, SAF has changed with the times, adding and then subtracting publications to its roster over the years. Of the 25 papers included in this anthology, 23 were originally published in the *Journal of Forestry*. The other two were published in *Forest Science*. Notably missing are papers sourced from the *Northern, Southern,* and *Western Journals of Applied Forestry*. This was not overt. The *Journal of Forestry* simply housed more papers directly related to fire. As well, it reaches back further into time as the applied journals were not launched until 1984, and ceased publishing in 2014.

In a collection like this, appreciation for the selected seminal papers relating to the American wildfire scene—which spans more than a century—can be easily be overshadowed by the observation of what is missing. There *are* important papers that have been omitted from this book, and while those that *have* been selected by no means represent an exhaustive list, they are exceptional and contribute significantly toward telling the collective story and helping us to contemplate the future.

Wildland fire managers make order out of chaos. Approaching how to structure this anthology was really no different. The objective of this anthology, as briefed to me by

Jennifer Kuhn, SAF's outstanding Director of Publications, was to identify the "seminal papers" from the publication history of SAF. Like fire, the task sounded simple in theory, but was complicated in practice. A seminal paper, to me, is one that has the potential to influence later developments and practices. It encompasses a person, a moment, or a contribution. Fortunately, the source material from SAF that encompasses these attributes is rich.

Wildland fire managers succeed by making a large task smaller through division. Thus, the idea of dividing up the subject into topic areas seemed natural. However, given the breadth of possibility, this was another daunting scenario. We ended up settling on the five topic areas that are included herein: fire policy, fire management, fuels management, a looking forward while looking back, and the utilization of fire. We could easily have doubled that to include fire prevention, technology development, aircraft utilization, injuries and fatalities, public information, organizational history, pure science, and so on. That task may fall to the next editor fortunate enough to be tasked with assembling the next anthology for SAF.

I won't go into too much detail about these five topic areas, but I implore the readers of this anthology to take the time to read the insightful introductions to each topic. All of these original pieces are stunning jewels in this crown of writings. However, I'd like to indulge some about the process and philosophy for making these divisions.

Section 1 provides an overview for the policy of fire management. All action on the ground—the tangible aspects of any profession—has its roots in policy, and new policy decisions set the tone for the future. Where would we be had the 10 a.m. policy not been instituted? Perhaps this is personal bias, but I believe that policy is the most important facet of wildland fire management.

It appears that identifying what is wrong with policy may be easier to determine than what right could look like in the future. A thoughtful introduction is provided by Dr. Ross Gorte.

Section 2, introduced by Dr. Elizabeth Reinhardt, acknowledges the most fundamental fact of American fire management: that we are concertedly attempting to manage it. This topic area stems from a belief that statistics and analysis will illuminate the way toward efficient control, to an inward look at how much risk we wish to assume to achieve this goal. This topic area especially could have included a great many more excellent articles.

Brian Van Winkle wrote the introduction for Section 3. Fuels management has always been a contentious issue, and despite the best efforts of so many men and women, perhaps the unfulfilled and not fully supported promise to solve the wildland fire problem is worth revisiting. The art and science of fuels management has been inked with the blood, sweat, and tears of the folks professionally involved with it. What the future holds for this endeavor is unclear, but the papers included in this part illustrate a sure reflection of how we have evolved. Notably, this part on fuels concludes with what is the most contemporary paper of the anthology, a paper from 2012 authored by three of the most influential modern voices. It also includes what is arguably my personal favorite of the whole lot, written by the legendary Aldo Leopold in 1924.

Section 4, with an introduction by Dr. Alistair Smith, includes papers that were unable to be fit neatly into one of the other specific parts. Gisborne's *Mileposts* paper contains some of the most powerful quotes in this book, and is itself an analogy for the structure of this anthology. You may recognize his famous scathing indictment of the 10 a.m. policy, as it is one of the most beautiful, devastating, and insightful lines to be published about wildland fire management and its unintended consequences. Harold Weaver's iconoclastic work on prescribed fire is important beyond measure. The role of pre-European anthropogenic fire is a subject area unto itself, as is the 1988 Yellowstone debacle. Rounding this topic out is a unique article forwarded by another professional organization, making a plea that is as of yet largely unanswered, but one that should not be forgotten.

Section 5, introduced by Kelly Martin, touches on the utilization of fire and finishes the anthology. It is the mirrored companion to Section 2. If there is any division at all among fire professionals, it resides in those who wish to fight it and those who wish to light it (or allow that which is naturally lit to burn). More, perhaps, have sided with the suppression mantra over the years, however many also have tried to keep the "vestigial flame," as Stephen Pyne puts it, alive within the profession. Stoddard's paper was an early rebuttal to the notion that fire had no place on the landscape. Ignored as this position was nationally, it helped maintain the southern fire insurrection. From a historical standpoint alone, Conarro's paper is worth a read, especially if you ever wondered how the term "prescribed fire" was coined and described. The scientific heavyweight H.H. Chapman defends the use of prescribed fire, in this case for longleaf pine sustainability. Cooper offers implementation strategies and proffers that perhaps there *is* a role for prescribed fire outside of the South. In 1975, the same year as Cooper's paper, the standards for use of fire in the West were being actively devised; this is reflected in Kilgore and Sando's paper about fire in the Sequoias. Perhaps my only real disappointment for inclusion is that Harold Biswell never published in one of SAF's publications (hence the exclusion of his work), although this last paper was written by some of his mentees and serves that important role.

One last note on how the papers were selected. You may have noticed that there aren't too many new papers as of the time of this publication. I felt that in order to truly meet the intent of selecting seminal papers, there must be a requisite cooling off period in order for the importance of a particular paper to be revealed. The average age of all the papers is 56 years, placing the average publication year at 1961, even when excluding the oldest and newest papers as outliers. Perhaps this is no coincidence, as 1961 is placed squarely in the golden era of fire research.

This anthology only hints at the depth of the well that the archives of SAF's journals hold for research and commentary. When going through the process of selection, I was pleased to discover that the papers included had a way of making themselves be heard. They seemed to whisper to me in some way that perhaps they needed to be listened to again. I would love to think that what they were saying was that their message had not been fully integrated and that their lessons are still there to be learned. I urge you, the reader, to see how these papers speak to you, and see if there is anything hidden inside that is inspiring, as many of these papers have been to others over the years.

Acknowledgments

Being invited to curate this anthology was an immeasurable personal honor and a task that was both exciting and daunting for the reasons described in my introduction. Having the opportunity for acknowledgments, I will indulge. I am compelled to acknowledge the people who influenced what became the direction of this anthology. I'd like to acknowledge the work that Jennifer Kuhn, SAF's director of publications, contributed to this anthology. She helped keep all of us who participated in line and on time, provided critical feedback and direction, and overall enabled the book to happen. She was a pleasure to work with and a true professional. I also thank SAF's Assistant Managing Editor Linda Edelman for copy editing portions of this work.

My old friend Brian Van Winkle is the one who forwarded my name to SAF for this task, and for that I thank him. His reward was to get roped directly back into it as an author to introduce Section 3. We are indebted to all of the authors who contributed outstanding topic area summations. Some of these folks were pulled out of retirement for the task, and we deeply appreciate all of them for volunteering to set time aside from their busy lives to work on their narratives.

Fire on the Land is made so much the better for their contributions.

Dr. John Groninger first provided mentorship in forestry while schooling, and that mentorship continues to this day. Dr. Charles Ruffner then ruined it all by starting a wildland fire crew while I was in graduate school; I'm a proud plank owner of the FireDawgs, a group that changed the direction of my career dramatically. However, I maintain that fuels management is just another aspect to forestry. Soon after starting my career, Dr. Jason Greenlee also influenced the direction of my work by gifting a dozen or so boxes filled with duplicate journal articles from his extensive private collection. I fell in love with all the old papers, and that love and understanding of history is the foundation of how I approached the work of building this anthology.

And finally, to by best friend and wife, Amanda, who from the very beginning has tolerated me coming home smelling like smoke, being away too often, and allowing all those boxes of papers in the garage to linger. One of the greatest gifts I ever received, besides her love, were the six books about fire she bought for me upon graduation all those years ago.

Part I

As Solid as Smoke: Wildfire Policy

■ **Dr. Ross Gorte**

This section of *Fire on the Land* contains five papers that illustrate the evolution of wildfire policy over the past century. Readers will notice that throughout this anthology, articles from the Society of American Foresters' journal archives are presented in chronological order based on year of publication. There is one exception. The first paper, by H.T. Gisborne, is not the first chronologically, but it *is* a useful opening paper nonetheless. Gisborne focuses on the question of whether foresters are, and should be, tree specialists or land management generalists, and asks implicitly if trees are the most important resource on wildlands. These issues were debated

when I was a student, and to some extent they continue to be discussed today.

Gisborne, as a prolific author on fire and other issues, addresses wildfire in a significant footnote. He indicates that fire control is a "service division" that supports management of all natural resources; this upholds his argument for unified natural resource management, rather than fragmented, resource-specific management. His point is that fire supports resource management, but is not a primary land management purpose in and of itself. This dominance of fire control within the field forestry generally and the Forest Service specifically has been an issue since the Big Burn of Montana and Idaho in 1910, and the concept has re-emerged in the past decade.*

"The New York Forest Fire Law," published in the *Journal of Forestry* in 1903, is the second article selected for this section. It focuses on the need for fire suppression, the duties of the various participants in fire suppression, and the need to regulate the careless response to fire. The emphasis on fire prevention and control were apparent at the time and continue to be a focus of wildfire management today.

The third article is by Irvin T. Haig. He notes the "very tangible" evidence of fire damage to timber and forage, and then discusses soil erosion and the damage to water quality and quantity. He concludes that the "importance of fire control in modern forest management can hardly be overemphasized." Later, Haig notes the variability of wildfire effects on soil fertility, and suggests that categorical statements on these effects can only be reached for differing conditions. Haig then discusses the inattention to the possible use of fire for disposing slash, for clearing firebreaks, for controlling pests, and for preparing seedbeds for regeneration.

The inattention is likely due to the Forest Service's long history of aggressive fire suppression. This view of fire was prompted by the 1910 fires in the northern Rockies. It was first institutionalized in the 10-acre policy in 1916; this policy stated that all wildfires should be contained to 10 acres or fewer. Aggressive fire suppression was furthered when a meeting of regional foresters in 1935 approved the 10 a.m. policy; this policy directed that all fires be contained by the onset of the next burning period: 10:00 a.m. Furthermore, light burning (prescribed fire) was castigated by Forest Service's

then-Chief William Greeley as "Paiute forestry" with no place in modern practice.

In this section's fourth article ("Fire Management Policy in the National Forests —A New Era"), Thomas Nelson described the emphasis on such aggressive fire control, and its attendant cost, as being inconsistent with integrated land management planning required by the recently enacted National Forest Management Act of 1976. He acknowledged that aggressive suppression was probably appropriate in the early years of the Forest Service, but that times had changed. As such, he described a new fire policy to supplant the 10 a.m. policy. The new policy had three main goals: to provide cost-effective control balanced with threats to life and property; to use prescribed fires to protect and enhance natural resources; and to integrate fire control and use with land management planning. This new fire management policy would, at times, lead to fire control efforts that were less aggressive than under the previous policy.

The last article is by Steelman and Burke ("Is Wildfire Policy in the United States Sustainable?"). They describe the current wildfire problem: significantly greater acreage burned than in previous decades. They note that the rising burned acreage is widely ascribed to higher fuel loads, resulting from changes in land use patterns over the past century, including logging practices, overgrazing, and aggressive fire control efforts. This situation has been exacerbated by climatic changes, including the extended drought in much of the intermountain west and the southeast. Steelman and Burke discuss the several policy changes that have been instituted to address the problem, including the National Fire Plan (2002), which commenced under President

---

*   For more on the historic view, see Stephen J. Pyne, *Fire in America: A Cultural History of Wildland and Rural Fire*, Princeton University Press, 1982. For a particular view on the more recent debate, see Robert H. Nelson, *A Burning Issue: A Case for Abolishing the U.S. Forest Service*, Rowman & Littlefield Publishers, n.d.

Clinton's administration, the 2003 Healthy Forests Initiative of President George W. Bush's administration, and the Healthy Forests Restoration Act (2003). These actions focused on changing the process to reduce wildfires through fuel reduction and ecosystem restoration, using more collaboration. They suggest that these policy changes have been less effective than they could have been because of the continued increases in the already high fire suppression expenditures, the stagnancy (not increase) in funding for fuel reduction, and the relative lack of funding and emphasis on ecosystem restoration and collaboration. Steelman and Burke suggest that a shift in priorities is needed to implement the policy changes of the National Fire Plan, Healthy Forests Initiative, and Healthy Forests Restoration Act, and thus to achieve a more sustainable wildfire policy.

The articles included here in *Fire on the Land* trace the evolution of wildfire policy over the past century and more. From the outset, policy has emphasized wildfire suppression, especially for fires ignited by careless users. Policy has also acknowledged wildfire use in many circumstances. Although initial views were vigorously opposed to prescribed burning, "Paiute forestry" and the nature and extent of the circumstances where fire use is seen as appropriate have expanded.

In the late 1970s, wildfire policy was modified to focus on resource goals established through land management planning and to constrain high wildfire suppression costs. This effort was probably doomed to failure by the stochastic nature of catastrophic wildfires and of the weather patterns that drive those wildfires.

The increase in wildfire burned area in the 21st Century has led to several initiatives to constrain wildfires by expanding fuel reduction and ecosystem restoration through greater collaboration among the various interests. These changes may also be doomed by the persistence of aggressive fire suppression with little regard to the cost and effectiveness of such efforts, and by the authorization for fire control to take funds from other resource management accounts, including fuel reduction and ecosystem restoration, without regard for the long-term consequences of such actions. Support for this current approach will not likely change until the fire community, as well as local communities and politicians at all levels, understand and acknowledge that, when driven by wind and drought, wildfires are uncontrollable natural events akin to hurricanes and volcanic eruptions.

The Challenge to the Society of American Foresters

1943. *Journal of Forestry*. Vol 41. No. 11.[1]

H.T. Gisborne
Senior silviculturist, in charge, Division of Forest Management, Northern Rocky Mountain Forest and Range Experiment Station, Missoula, Mont.[2]

Abstract: The place of multiple use in the management of wildlands in the West has been described and emphasized in the eight preceding papers originally intended for presentation at the postponed 1942 annual meeting of the Society of American Foresters. It is fitting that this last paper of the series the implications of multiple use to the Society should be considered. The author points out that professional foresters must devote time, energy, and thought not only to timber production and management, but also to such other resources of wildlands as water, forage wildlife, and recreation. He believes that the management of all these resources falls within the province of the forester, irrespective of whether or not trees are present on the lands involved, and therefore urges that the Society of American Foresters interpret "forestry" as synonymous with "wildland management." Such action would place forestry along side of agriculture and engineering as one of the three great land professions, and would enable the profession to play its rightful part in postwar programs directed at the conservation and improvement of our natural resources.

The major topic which was to have been discussed at the 1942 annual meeting of the ,Society of American Foresters was "Multiple Resources of Wild Lands in the Mountain and lntermountain Regions of the United States." Its basic economic significance and the truly national importance are clearly evident in the caption of the first of the three phases into which the program was divided: "All. wildland resources are important as an integral part of the general economy of this

---

1. Prepared for the postponed 1942 annual meeting of the Society of American Foresters.
2. About background for this article, Mr. Gisborne writes: "Fire control is not one of the fundamental arts or sciences of forestry; it is a '·service' division which should distribute its aid equitably among the basic production divisions, i.e., timber, range, wildlife, recreation, and water. But fire control cannot function equitably and efficiently until the basic divisions are integrated into one clearly unified field of effort. The present article is intended, therefore, to help solve the biggest problem confronting both fire control and forestry."

area and of the nation." If this premise is accepted, and it certainly should be after study of the papers now published, then one great question looms as the real, the paramount challenge to our Society: *Is forestry wildland resource management, or is forestry confined to trees?* Until we answer that question, there is no point whatever in discussing requirements for membership in the Society of American Foresters. If we decide that forestry is primarily tree growing with all other wildland products subservient to trees, then we shall have a definite basis upon which to specify membership requirements. If forestry is wildland management however, with timber dominant here, forage there, recreation, wildlife, or water elsewhere, we shall have a quite different basis.

Forestry Is at an Open Door

Furthermore, there is urgency in this challenge. This is true not only because of present pressures created by the war but because after the war tremendous quantities of labor and millions, perhaps billions, of dollars will be directed at the conservation and improvement of our natural resources-soil, water, and minerals. General plans have .already been published, particularly by David Cushman Coyle (3, 4) and by Morris L. Cooke (2). Additional, more specific, and official plans are known to be in the making. When the time comes to put those plans into effect, professional foresters will be expected to assume their share of this great responsibility.

What is the job then to be allotted to the profession of forestry? Is it tree-growing alone, or is it wildland management? If the· Society of American Foresters represents the profession, it should be ready to help in the formulation and execution of those plans. It should be ready first with a clear concept of forestry and its place in land use, all land use; and it should be ready before the urgent need of immediate application overwhelms us like a flood.

In this approaching need, forestry is not "at the crossroads." It is instead standing before an open door. The question is: Shall we shut the door and stay inside our little "treehouse" or step outside and take our place in the bigger world of wildland management, shoulder to shoulder with the managers of cultivated land and side by side with those most intensive users of land, the construction engineers? There seems to be a place and a need out there for someone to plug the present apparent gap.

"The Old World Is Dead"

Recently an acknowledged authority on world events, Edward H. Carr, chief editorial writer of the *London Times*, made the following statement: "The old world is dead. The future lies with those who can resolutely turn their backs on it and face the new world with understanding, courage, and imagination." These words may be brushed aside as mere words; they may be admitted as applicable to international affairs, "big business," etc., but not to forestry; or they may be accepted and used as a profound and prophetic statement.

In a recent number of the *Industrial, Bulletin*, issued by Arthur D. Little, Inc., industrial research consultants, the statement is made that "perhaps one of the most basic changes wrought by the conversion of industry to war production is stimulation of the tendency to think of a company as a management unit, rather than as a collection of physical equipment. As a management unit, the company's field of operation becomes far less specialized than would be indicated by the company's physical situation ... in all conversion, however slight, the management's thinking is broadened beyond what it had formerly considered its natural limits." Are we broad enough to recognize forestry as a distinctive management unit dealing with wildlands as a whole, and not as a mere collection of specialties dealing only with isolated segments of the problem?

Possibly forestry will not, or should not, let itself be affected by the present world-wide upheaval in social, economic, and industrial affairs. Possibly the new methods of manufacturing useful products from cellulose and lignin will have no effect on the use of land for the growing of trees. Possibly the new world demand for beef and mutton, leather and wool will have no effect upon the use of wildland to grow these products rather than to grow cellulose, lignin, and sawlogs. Possibly the new social life with more people traveling by rail, automobile, and airplane will have no effect upon the competitive demand for wildland dedicated to recreation rather than to livestock or timber production.

These are matters of opinion, but the Society of American Foresters might do well to form and to express an opinion. If the old world is dead and the future lies with those who can resolutely turn their backs upon it, the Society of American Foresters and the profession of forestry may be vitally affected. We might do well to remember that progress requires change. You cannot go ahead anywhere or in anything without changing your position. Before we move, however, we should try to be sure that when we do move we shall go forward, not backward.

Real progress involving real change does not, of course, require expansion. A destroyer does not have to expand into a cruiser or a cruiser into a first-class battleship in order to reach its destination or to accomplish its purpose. Similarly, tree forestry undoubtedly can adjust itself to the new economic conditions and can drive straight ahead toward its fine objective of more and better trees for the greatest good of the greatest number in the long run. Without doubt, it can remain as treeland forestry and still make progress.

But even if it confines itself to the management of land with one or more trees on or near it, forestry never has been, is not, and cannot become pure silviculture. Our own history tells us that the real, universally accepted "forest" is and always has been an inseparable trinity of trees and brush and grass. It is the function of the forester to manage all three of these products of wildland. Not just one of them, but all three of them.[3] If this be true, on what basis is the forester entitled to claim that the technically trained man who manages wildland having two of these crops, brush and grass, but not a tree, is not a forester? What if that technician is managing wildland having a cover mostly of brush and grass, and with a few noncommercial trees? Is he to be admitted to the fold? Or must he have one or more trees that will make sawlogs?

Wildland Management the Big Problem

Foresters should not quibble over such small distinctions while the great field of wildland management goes begging for technical mastery. Foresters are already managing the bulk of those wildlands. Most of those foresters are silviculturally trained. Many of them have needed and used other training more than silviculture. But all of them are doing one distinctive

---

3. Schools of forestry offer separate majors in timber management (trees), wildlife management (brush), and range management (grass). The major in timber management alone is called "forestry." Is there a school in the country offering a course giving a student the principles required for determining which parts of an area should be used for timber, which for brush, which for grass, and *why*, in each case? Such a course, in forest management for any or all possible uses, not trees alone, is conspicuous by its absence in the training of foresters. It is conspicuous also by its essentiality on the actual job of determining forest land use for the greatest good of the greatest number in the long run. There is reason for doubt that most of the schools are teaching anything more than isolated techniques, even for treeland management.

job, different from engineering, different from agriculture-managing wildlands. Progress challenges us to claim this field or to relinquish it and specialize on treeland alone. Progress also challenges us to heed the words of David Cushman Coyle: "There must be formed in the public mind an ideal to which all programs are related and by which they will be judged." Our program must deal with trees alone or with wildland management.

In meeting this great challenge, the Society of American Foresters is the conglomerate mind and the concerted voice of the foresters of the United States. The great and influential public which we serve knows this. How do you sup· pose it personifies us? As primarily firefighters, or timber growers, or lumberjacks? As game wardens here and sheepherders there, or as glorified rangers with lipsticky Paulettes pouring tea from silver services in our ranger stations while we parachute out of our airplanes?

The matter is no joke because it is a human characteristic to personify a profession or a well-known group of men. We are all so familiar with John Q. Public and with Mr. Average Congressmen that we recognize both of them on first sight in our highly effective American cartoons. We might do well to consider the personification which our Society deserves. I know that I should like to see the forester standing squarely beside the multi-purposed engineer and the diversified farmer as an all-round wildland manager with a specific, economic plan and program in his hand, not merely a nurseryman-lumberjack, tree forester, waiting to be fitted into one small niche of somebody else's plan.

The Forester Already a Wildland Manager

To many foresters this may seem to be extending the term "forester" almost as broadly as Professor Jackson extends the term "engineer";[4] or as some ecologists extend "ecology" until it encompasses the whole human race and its cosmic environment. But the history of forestry shows that the concept of a forester as a wildland manager; practicing multiple use, is not a novelty and not an exaggeration. Foresters were game conservationists and livestock producers long before they were timber growers. As recently as 1914, Pinchot (6) believed that "forestry is the art of producing from the forest [which he did not define] whatever it can yield for the service of man."

Actually, we ourselves have perverted the term "forest" until it no longer means primarily a tract of uncultivated land, but is now frequently restricted to merely the dense stand of trees upon that land. The compilers of dictionaries naturally have followed our lead. Possibly this perversion has already gone so far that it will be impossible to rescue the original and full meaning of "forest." We are now in dire need of just such a term and might remember that we still have the word "silva" to signify the stands of trees cultured by the silviculturists, as well as "timber" which probably is the most usable designation. But timber culture is not all of forestry or wildland management.

To some soil conservationists the distinctive category of "wildland manager" may seem to be made to their order. Undoubtedly they could fill it. But they could not stop there, for if they are true to their name and awake to their opportunities, their field of

---

4. The author does not use the terms engineer· and engineering as broadly as Prof. Dugald C. Jackson who, in an address entitled "Engineering's Part in the Development of Civilization" (Science 89:235), defined engineering as "a process of planning, organizing, ·and executing work concerned with directing the forces of nature to the service of mankind."

effort goes far beyond forests or wildlands. In fact, their major field of effort and their most valuable contributions to socio-economics undoubtedly lie in the rich agricultural lands of the farmer. Furthermore, at times the soil conservationists will even step far over into engineering, that most intensive of all land uses, to stabilize road shoulders, bind the surface of aviation fields, prevent the sedimentation of dams, etc.

But their work with the farmer does not make the soil conservationists farmers, nor does their work on roadsides and airfields make them construction engineers. Their little work, probably the least required of all, in treating wildland soils, similarly does not make them foresters. Instead, the soil conservationist is clearly a broad specialist with a very definite function in each of the three major fields of land use-forestry, agriculture, and engineering. He is "doctor" to all three of these basic, economically different professions.

Three Great Land Professions

Viewed thus, from the doorway of our forestry "tree-house," there are three great professions dealing with land-the basis of our civilization. At present the engineers are in the foreground accomplishing miracles by expropriating to their intensive use whatever lands they need. Pasture, wheat land, orchard, or celery patch-it makes no difference if a landing field is needed at that spot. The richest loam in the most accessible place, blessed with the best climate on earth, will be covered with cement and a roof put over it if a factory is needed right there. Or a mine shaft will be sunk and roads built to it, regardless of other possible land uses, if that ore will serve a national need. This is right. It is good economics. It is "for the greatest good of the greatest number in the long run."

Just behind the engineer the agriculturist also labors to this same end. But he has other objectives, in addition to supplying the current demands of our nation and others. He must not destroy the soil from which his future crops must come, for what is a nation profited if it shall gain the whole world but lose its own soil? The farmer, the agriculturist, cannot overlook this. He must cultivate with care for much more than this year's crop. To meet the demand he may have to cultivate more intensively than ever before, and he may even be forced to reach out and take for himself and his uses land never before cultivated. That means take from-whom? The forester. For who else is there working "out there" in the great field of land management, maintaining a source of land ready when needed to flow from no cultivation into cultivation, and ultimately into dams and roads, factories and airfields? What profession should you expect to be working out there backing up the farmer, backing up the engineer, protecting the original lands that God gave us and upon which all users must draw? Certainly we and the public should visualize someone.

Actually there is no "gap" in the ranks of those working "out there." The forester is already there. He is there in the Soil Conservation Service, the Grazing Service, the Indian Service, the Park Service, the State Service, the Forest Service, managing our reservoir of wildland, keeping those lands as productive as possible of whatever they can yield for the service of man, without intensive cultivation. That is also approximately the case for the private foresters; they manage the best timberlands in the nation almost exclusively for timber production because of the national and even international need for this one forest product-wood. But they too are selling livestock forage; they even permit the public to use their privately owned and expensively protected lands for public hunting and fishing. They do this last, of course, largely because the public still holds firmly to the original concept of a forest as "a tract of woodland and waste, usually belonging to the sovereign, set apart for game." The major change in this concept is that the sovereign's forest, or at least his right to hunt, has become the people's right.

Hence the forester is actually already on this basic job and he is doing far more than merely raising trees. He has taken his wild-land place in our national and even world-wide economy, with the rich brown mulch of the forest on his hands and his sleeves rolled up, contributing to one of the three great shirt-sleeved professions—engineering, agriculture, and forestry. While the three "learned professions," the ministry, medicine, and law, may look somewhat slantwise at their shirt-sleeved brethren, they themselves would not long survive by their sole efforts. Likewise, if the engineers and agriculturists are to have new fields to conquer and if they are to have someone make use of their discarded lands, they too must look to another profession, the wild-land managers—the foresters.

The Paramount Challenge

And here lies the paramount challenge to our Society of American Foresters; Are we these essential wildland managers or are we only one little segment of that great profession? I contend that foresters are now doing the job, with varying degrees of success, on public and private lands alike. I suggest that we get together and admit that that's where we are and why we are. If so, then let our Society speak henceforth as a society of American professional men engaged in all that the term "forester" ever has implied, in all that it can imply for the benefit of civilization.

Most certainly the Society of American Foresters should have something to say, something helpful to contribute in the national soil and water conservation plan which Cooke (2) evaluates as a 3-billion-dollar job. Most certainly the Society should be able to help on the 945-million-dollar part devoted to "lands not suitable for cultivation, and which should be retired," and on the 400-million-dollar portion concerned with "lands in farms, not in cultivation." These are wildland, "forest" problems if cultivation, i.e., . agriculture, is "out."

Silvicultural practices may solve the problem for part of the 575 million acres involved, but for only part, and probably a small part. Grass and brush are the more likely cover crops on most of it, with livestock, wildlife, and more efficient water control the principal products. But for every acre, "the maximum yield of useful products is obtained, as a rule, not by making all other forms of life subservient to a single one, but by building up a balanced community of living forms, each one of which contributes its share toward the area's total yield," to quote Cooke.

This is multiple use of wildland. This is the forestry about which our present symposium is concerned. This is the forestry outlined by Marsh and Gibbons (5), but concerning which their "action program" concentrated exclusively on timber, as if watershed, recreation, wildlife, forage, and other uses would take care of themselves. They used the word "forest" to include all the crops and benefits, but they used the word "forestry" as Professor Chapman did at Jacksonville (1), as synonymous with silviculture and timber management only. These are vital matters in which the Society of American Foresters is interested, concerning which it can help, and on which it should speak.

Similarly there are David Cushman Coyle's programs as outlined in his book *Roads to a New America* (3) and his booklet *Our Forests* (4). Coyle brings out the problem, or the opportunity, depending upon your viewpoint, of the revival of labor sources similar to the C.C.C. and the W.P.A. Of the wildland share of these socio-economic investments· to be made after the war, what proportion of such labor should go to timber, what to range, what to recreation, what to water control? Is our conglomerate mind and concerted voice ready, with a specific program?

The Society Should Lead

The Society of American Foresters should not sit back and wait until after the war before expressing itself on such questions of major policy which are sure to spring up or be suddenly sprung upon us then. We should not sit back and depend upon engineers like Cooke and Coyle to furnish the essential over-all plans, even though Cooke and Coyle may be the topmost leaders of thought in their profession. Our Society can and should examine the various plans submitted by all such authorities, lend our support to the good features, amplify them wherever possible, oppose those which we believe to be bad for our land or inimical to the greatest good of the greatest number in the long run. We can help. we can help most as a society of professional foresters dealing with all aspects of the wildland problem.

There are only three choices: (1) We can slam the door and stay in our silvicultural tree-house; (2) we can leave the door open and stand in the doorway, making gestures; (3) we can step out and take our rightful place alongside the engineers and the agriculturists in the big job ahead. Mr. Carr suggests the use of "understanding, courage, and imagination."

Literature Cited

1. Chapman, H. H. 1942. The problem of curricula in institutions giving professional instruction in forestry. Jour. Forestry 40:182–189.

2. Cooke, M. L. 1941. On total conservation. In Univ. Penn. Bicentennial Conference Conservation of Renewable Natural Resources: 149-159. University of Pennsylvania Press, Philadelphia.

3. Coyle, D. C. 1938. Roads to a new America. Little, Brown and Co., Boston.

4. ———. 1940. Our forests. National Home Library Foundation, Washington, D. C.

5. Marsh, R. E. and W. H. Gibbons. 1940. Forest-resource conservation. U. S. Dept Agric. Yearbook 1940: 458–488.

6. Pinchot, G. 1914. The training of a forester. J. B. Lippincott Co.; Philadelphia.

Fire in Modern Forest Management

1938. *Journal of Forestry*. Vol 36. No 10. 1045–1051.

Irvin T. Haig
U.S. Forest Service[1]

Few foresters in America would quarrel with the statement that "the most remorseless associate of man in the destruction of native vegetation is fire." Indeed within the general limits fixed by climate and soil no other natural phenomenon has left its mark so unmistakably upon the vegetation of North America. Early travelers (Trumball, Batram, Nuttall, Lyall) along the heavily wooded eastern coast reported few clearings "except where timber destroyed and its growth prevented by frequent fires" but everywhere occurred extensive areas of subclimax associations almost certainly of fire origin.

Indeed almost every forest climax of North America displays a fire subclimax, usually occurring over a considerable area. In the northern forests of spruce-fir these fire climaxes are chiefly the jack pine or aspen-birch types. Pure white pine stands, however, were undoubtedly far more extensive than they would otherwise have been through the action of this agency. In the Piedmont region and along the southern coastal plain frequent fires have maintained a pine forest over a vast area and retarded the development of a broadleaf climax. Further west along the borders of the mesophytic hardwood associations fire had maintained a subclimax of prairie nature. Many ecologists believe that in this same region the common oak-hickory stands represent a fire subclimax of the maple-beech type. In the West such important types as Douglas fir, western white pine, and lodgepole pine have been perpetuated in more or less pure form by fire, Sears calls attention to the fact that in analyzing the peat deposits in the Medicine Bow Mountains of Wyoming it is evident that fire has been a direct agent in causing an alternation between lodgepole pine and spruce-fir. Some of the peat deposits examined carry a story of fire going back more than 10 or 15 centuries, clear evidence of the work of fire long before the days of Columbus. Extensive stands in California of such species as knobcone pine

---

1. Thursday Morning Session, June 30, 1938. Joint Meeting of Forestry Groups and the Ecological Society of America. *Chairman*: C. C. Adams. *Chairman Adams:* The Program Committee has arranged a very interesting symposium on the influence of fire on forests, wildlife, and public welfare. I am certain the members of all the various groups represented here today have a common interest in the subjects to be discussed this morning. The Chairman then called for the presentation of the following papers.

and Monterey pine and the brushfields of the pine region show the effect of this same agent.

This situation has been greatly aggravated of course with the advent of lumbering and fire combined. As a result extensive areas of such associations as aspen-birch now form a common feature of the landscape in the Lake States while the brushfields of California in the pine region are now estimated to total more than 2 million acres in area. These are merely cited as two examples of extensive similar phenomena.

Fire Damage

In similar manner many aspects of direct fire damage are so obvious as to war- rant but mention. Most of us have seen the snag patches or "brush" fields that show the hopeless devastation caused by fire or by a combination of logging and repeated fires; and something of the more subtle if none the less important losses in trees and tree defect introduced by fire in less spectacular form. "He who runs may read" in North America the story of tangible damage etched by fire into our timber resource.

Small wonder then that fire control operations loom as one of the major tasks in modern forest management; that foresters have bent every effort to reduce fire losses to within acceptable bounds and have properly regarded this step as fundamental to the successful practice of forestry in America. It is doubtful if in any other country in the world, fire protection needs have been given the intensive thought and study which they have received here. But even on the national forests where average losses approximate the permissible area burnt, critical areas continue to show the need for more intensive effort; while on land poorly protected or not under organized protection, the actual losses are still over ten times as great as they could be if the area burned annually were reduced to acceptable limits.

The very tangible losses in timber and forage may well be more than matched by the indirect losses upon which foresters are only now beginning to accumulate a respectable amount of factual data. A tidy body of accumulating evidence from many sources and conditions shows beyond all reasonable doubt that destruction of vegetation, by fire or otherwise, results almost invariably in increased surface run-off and consequently in property destruction, as in silting of reservoirs and greater damage to other stream improvements, and in the usability of the water and in recreational values, entirely aside from the effects on magnitude and regularity of streamflow, and the losses in soil fertility due to such washing, irreparable except in terms of centuries, are far too important to be ignored. Many of these facts are readily demonstrable. Bennett reports paired plots in the post oak type, one burned, the other with natural litter from which the run-off during a one-month period of continuous rainfall totaled 250 gallons of clear water per acre from the unburned plot and 27,600 gallons of muddy water per acre from the burned plot. The soil eroded from the burned plot was 15 times as great. In the pine region of the Sierras a five year record from repeatedly burned and comparable unburned plots shows a yearly run-off from the burned area ranging from 31 to 463 times that of the unburned with a yearly erosion ranging from 2 to 239 times that from the unburned. Many other studies show the increased surface run-off and erosion from barren lands under conditions similar to these following severe fires. (Meginnis, Bates, Munns, and Sims). More spectacular examples are afforded by such events as the disastrous flood in southern California on January 1, 1934, which caused the loss of 34 lives and millions of dollars in agricultural and residential property. The watershed in which the destructive flood originated had burned over only a few weeks earlier. Neighboring watersheds with their forest cover intact yielded clear water which caused no unusual erosion and did little damage. The San Dimas, for example, had a flow of only 50 second-feet per square mile and a load of only 56 cubic yards, while the flood discharge from the burned drainage basin reached a maximum of 1,100 second-feet per square mile and carried some 76,000 cubic yards of eroded debris. There seems little room for doubt that although indirect damage from fire in terms of watershed values will vary tremendously by region and situation that losses from this source cannot be ignored in any fair appraisal of fire effects. The importance of fire control in modern forest management can hardly be overemphasized.

The Use of Fire

The problems of fire control have been so urgent that few foresters have stopped to think how extensively we use fire in current management practices. Fire is one of our most effective tools in the disposal of logging slash and the reduction of other special hazards. Controlled fires used extensively in the actual combat of fire and in the preparation for such combat in the clearing of firebreaks, roadside safety strips, and similar areas. Fire is a common agent in the control of insect pests and disease as in the burning of stems infested with bark beetles or Nectria cankers. In recent years fire has been employed in the preparation of planting sites in at least two important types, overmature western white pine types and California brushfields. The fact that these operations have been conducted successfully in two of the most dangerous of fire regions speaks well for the growing ability of experienced men to handle fire as a tool and portends more extensive use in skilled hands as confidence grows and is justified by success.

In the field of natural regeneration foresters in North America have been reluctant to accept the idea that under special circumstances fire may be a useful silvicultural tool. In the writer's opinion such hesitancy has been thoroughly justified in view of the appalling losses that might follow the use of fire in careless or unskilled hands and in view of the meager amount of acceptable data as to its possible usefulness. Accumulating evidence leaves little doubt, however, that fire does have a place under certain conditions in certain types. Hesselman has demonstrated the value of fire in the regeneration of both spruce and pine in northern Sweden wherever the humus is too heavy to be broken down by opening up of the crown canopy through felling operations. Not only is fire helpful in creating a more favorable seed bed, but it also aids in hastening the nutrification processes necessary to proper seedling development. Hesselman deplores, however, the use of fire where other means are available of initiating a satisfactory nutrification rate.

In North America accumulating evidence (Moore, Lowdermilk, Haig, Osborne, and Harper) indicates that in many coniferous types, soil bared by fire or other action offers, with some exceptions, a far more favorable seed bed than needle litter; and that fire, other conditions permitting, might be a useful tool in creating favorable conditions for germination. This seems to be almost solely due to its physical action in exposing mineral soil; thus furnishing a more efficient contact medium for supplying water to germinating seeds. Exposed mineral soil also dries more slowly and fluctuates less rapidly in both moisture content and temperature and over a narrower range than the humus layer, an important factor in the critical period of early germination and development when even temporary drying or passing exposure to unfavorable high or low temperatures may result in heavy seedling losses. The abundant mineral nutrients released by destruction of the humus layer may also aid in the initial development of the seedlings. Fire has also been used extensively in creating suitable regeneration conditions for chir, sal, deodar and teak (Troup, Makins, Hale, Smythies, and Unwin). In the chir pine (P. longi[olia) region as Gorrie and others point out, fire is used also as a routine protective measure, the fire hazard being reduced periodically by controlled fires on a two or three year rotation. Whether or not fire has such a place in similar types in the United States will remain to be seen. Show and Kotok emphatically reject the practicability of "light burning" as a management practice in the pine region of California because of the excessive damage involved, though they cite occasional benefits in the preparation of planting sites, the thinning of too dense reproduction and the killing of the less desirable fir with consequent improvement in stand composition from a commercial viewpoint. In the southern pine region the necessity of discouraging uncontrolled fires which annually do a tremendous amount of damage has made foresters reluctant to consider that fire has any place as a control measure even in highly resistant longleaf pine stands. Although evidence is accumulating on this phase and it is known that controlled fires, properly timed, can be run through pine stands without undue injury, the place of fire among other protection measures is still highly debatable.

Accumulating evidence does support the long-held opinion that fire is an important element in longleaf pine regeneration. Chapman has long advocated the use of controlled fires as a necessary element in longleaf pine regeneration arguing that total exclusion of fire means the extermination of longleaf pine as a type and pointing .out the usefulness of fire in seed bed preparation, in removal of competition from hardwoods, and in reduction of damage from brown spot

to a sufficient degree to enable longleaf to get out of the "grass stage" and make desirable initial height growth. Evidence now accumulating confirms these general conclusions (Osborne, and Harper, Wahlenberg, Green, et al, Siggers) and shows that controlled fire may be a useful and at times a necessary tool if adequate longleaf regeneration is to be obtained.

One of the most hotly debated points concerning the possible use of fire as a silvicultural tool revolves around its effect on soil fertility. Much evidence shows that burning almost inevitably results in a less fertile soil as far as physical qualities are concerned (Auten, Heyward and Barnette). But where the question of over-all fertility is concerned the situation is less clear. In general it seems to revolve around the question of whether or not losses in organic matter removed by fires will result in soil deterioration; one school of thought claiming that these losses will be more than compensated for by additional organic matter and nitrogen produced following fire. Various studies point out the reduction in potential nutrients by fire, but other studies show that these losses are primarily confined to humus layers and may be compensated for by the immediate liberation in the ash of a considerable quantity of available nutrients (Alway and others, Fowells and Stephenson, Isaacs). Additional compensation may follow through the neutralization of organic acids, the improvement of conditions for more rapid bacterial action, and the introduction of a soil building flora. Even repeated fires may not be seriously detrimental under certain conditions. Lutz found the soils of the pitch vine plains of southern. Jersey only slightly less fertile than nearby less frequently burned soils. Heyward and Barnette in one of the most comprehensive studies of which the writer knows, concluded that for the longleaf pine region soils burned frequently over periods up to 44 years failed to show any marked deterioration in fertility. Al- though the unburned soils were undoubtedly in better physical condition this was at least partly compensated for by slightly higher total nitrogen and organic con- tents and much higher calcium content in the burned soils.

In the light of existing evidence it seems doubtful if any categorical answer can be given to the question of fire and soil fertility and it will probably have to be worked out, as most complex biological problems must be, for each major condition involved. It seems doubtful under these circumstances if foresters are taking a sound or desirable position by citing fire as a soil destroyer where only direct action on fertility is con- cerned. No one would dispute the undesir- able effects of fire under other conditions where soil structure is destroyed, where burning is followed by excessive washing of the surface soil, or where the ability of the soil to absorb water is of paramount im- portance and this ability is impaired. Would it not be wise to place the need for fire pro- tection squarely on more tangible and less controversial items until the effects of fire on soil fertility can be sufficiently studied? The data on fire damage from indiscrim- inate burning are too strong to need such weak support.

Indeed in considering the whole question of fire use and effects it might be well for foresters to remind themselves of the nat- ural prejudice introduced by their rightful recognition of the need for more extensive and more effective protection from indis- criminate and destructive fires and approach the entire question in the scientific spirit so aptly described by Francis Bacon, i.e., with "a mind eager in search, patient of doubt, fond of meditation, slow to assert, ready to reconsider, ... not carried away either by love of novelty or admiration of antiquity and hating every kind of imposture, a mind therefore especially framed for the study and pursuit of truth."

COMMENTS
J. Miles Gibson, *University of New Brunswick*

I WISH to congratulate Dr. Haig on the excellence of his paper and the clarity he brings to this extremely interesting and frequently confusing subject. Instead of discussing the many interesting points brought out in the paper I will attempt to present a few of the phases of the effect of fire on Canadian forests.

As most of my personal experiences have been in the extreme East and the extreme West I will largely confine my remarks to conditions in two widely separated areas. My remarks will deal more with mainstand sequences than with all factors involved. The whole subject of fire effect on forests is filled with variables, many of which are ex- tremely difficult to measure quantitatively.

You first have a forest and then a fire. The forest may be one of a great many tree and plant associations and the fire may vary in intensity, in time of occurrence, and in the degree of burning. Actually many of these conditions may occur within the area of a single fire where variations in soil moisture, vegetation, and fuel may result in the differ- ence between a hot and light fire. This large number of variables certainly complicates the picture and one sometimes wonders that forests established after fire exhibit so many seemingly similar characteristics.

The forests of the West, as far as my ob- servations go, are set back for a longer peri- od as the result of fire than those of the East. This is probably partly due to shallow soil, making more vegetational sequences nec- essary before a tree crop is established and to the fact that many of the more severely burned areas having been logged previous to burning, the resultant fuel and high hazard intensifies the disturbance.

Fires in northern British Columbia result in lodgepole pine on the dryer benches and poorer soils and various age classes of this species often delimit the areas of separate fires. As the humus layer builds up, spruce and balsam may appear in such stands, but it may take a full rotation.

On fresh to moist areas with better soils, fireweed (Epilobium augustiJolium) is replaced by aspen, willow, birch, and fire cherry. The type or species association may be due to the severity of the fire and its effect on seedbed and moisture conditions.

Re-burns before the established species produce seed seriously retards recovery and causes a considerable lag in age of reproduction over the age of the burn. Delayed regeneration and hot fires resuit in a tendency to an open all-aged forest.

Gorman in investigating burned over lands in British Columbia indicates that the composition of the tree and plant community of distinct character is governed by soil type and available soil moisture. Intrusion of lodgepole pine is attributed to reduced soil quality and reflects the occurrence of severe burns which occurred before as well as after settlement took place.

Lack of satisfactory forests developing on areas severely burned is very noticeable in British Columbia and would be still more noticeable if some of the indigent species such as ponderosa pine and Douglas fir were not reasonably fire resistant when mature.

In the East and particularly in New Brunswick and in Nova Scotia many large forest areas have developed after severe fires. In the initial forest following the fires the more common associations are aspen, the birches (white and gray), and fire cherry which occurs particularly in the early stages. Associated with these intolerant hardwoods, one often finds an understory of red and white spruce, balsam fir and occasionally white pine. Frequently instead of intolerant hardwood conifer associations, red or white spruce and jack pine will occur either pure or in mixtures with white pine or more rarely with red pine.

The association may be the result of soil or moisture conditions, it may reflect the intensity of the fire, or it may reflect the past forest association. The association may frequently change abruptly; it may, on the other hand change gradually with no distinct transition.

On the poorer soils, particularly if poorly drained, black spruce becomes established while on the lighter, sandier soils jack pine becomes established; although in the latter case, aspen and white birch may develop in the initial stand.

In general spruce develops to a greater extent than balsam and one frequently finds forests in which white pine was formerly important succeeded by jack pine or jack pine and black spruce with a very limited occurrence of white pine.

A few days ago I observed an almost pure stand of black spruce on an area burned fifty years ago; although cedar, white pine, balsam, and black spruce comprised the original stand.

Southern and eastern New Brunswick forests are in many instances the result of past fires which destroyed the original forests. In general jack pine and white pine occur on the drier lighter soils, black spruce on the poorly drained soils, and aspen and white birch on the better soils with red spruce and balsam as an understory.

Associations of hemlock and tolerant hardwoods seldom reoccur following a fire and very frequently, although not always, tolerant hardwoods when burned are initially followed by intolerants.

Subsequent burns before stands reach maturity result in a much greater loss and repeated burns create conditions that will retard the possibility of a forest crop for generations.

The sequence towards a climax with one fire and no subsequent disturbance is not definite, although in New Brunswick the climax may be pure softwood, mixtures of softwoods and tolerant hardwoods, or tolerant hardwoods.

It will require at least one rotation for this climax to develop except in the case of the coniferous type. Hesselman's work indicates that controlled fire stimulates nitrification and encourages coniferous reproduction and in a number of places in New Brunswick the successional change due to intolerant hardwoods disappearing after acting as soil builders may be observed.

Some investigators indicate that white pine forests occurred as the result of a combination of fires and satisfactory seed years, but while white pine distinctly occurs on burned areas in New Brunswick it usually is only one component of the stand; although it may occur in pure groups. Cedar, either in pure stands or as scattered trees, may be observed on previously burned areas but the contradictory association of former cedar swamps developing into red maple or red maple and yellow birch associations after a fire may also be observed.

The results of fire as far as species association is concerned are so variable and the factors involved so numerous that one can readily understand why research in this field is necessary to give us the information which might assist us in using fire as a controlled agent to develop some of the favorable associations that have arisen from past fires.

In British Columbia I have observed decadent cedar and hemlock forest which I feel sure will never support a thrifty stand unless swept by fire. The same thing may be true to some degree where cut-over stands appear to be stagnated or consist of defective trees and brush which prevent the development of potentially valuable trees. While all burning involves a hazard it may be that the benefits to be derived might justify the hazard.

The New York Forest Fire Law

1903. *Journal of Forestry*. Vol 1. No. 4. 134–139.

 Clifford R. Pettis

It seems, on account of the disastrous forest fires which have recently occurred in the Northeastern United States, quite apropos to describe the fire law of the State of New York.

"Self preservation" is said to be "the first law of nature," and forest protection should be the first consideration in forestry. Fire is by far the greatest of all the destructive agents of the forest. A well-known writer uses the expression, "It takes thirty years to grow a tree and thirty seconds to cut it down and destroy it." Fire will destroy the thirty-year-old tree equally as quick and usually at the same time destroy the soil.

The law of the State of New York gives to the Forest, Fish and Game Commission the "care, control and supervision of the Forest Preserve,"[1] the enforcement of the laws relating to forest fires, and the authority to "make rules for the prevention of forest fires and cause the same to be posted in all proper places throughout the state."

The laws relating to forest fires consist of nine sections in Chapter 20, Laws of 1900, Article XIII. They may be summed up as follows: Section 220, Powers of Commission. Section 224a, Authorizing the office of Chief Fire Warden. Section 225, Town Fire Wardens and Fire Districts. Section 226, Duties of Fire Wardens. Section 227, Compensation of Fire Wardens and others employed at fires. Section 228, Railroads in forest lands. Section 229, Fires to clear land. Section 230, Forest fires prohibited. Section 231, Proceeds of actions for forest fires.

The wording and fundamental purpose of the law is to *prevent* fire, and at the same time have an efficient organization which can quickly get to the place and cope with a fire while it may yet be controlled, or render such services as may be expedient. The public welfare requires and the law provides that all forests, whether on state or private lands, be entitled to the same consideration and protection.

---

1. The Forest Preserve includes all the wild lands of the state situated in the Adirondack and Catskill counties, except the lands in the towns of Altoona and Dannemora, Clinton county, which are under the supervision of the Comptroller. The Forest Preserve includes all or portions of the following counties, of which 1,436,686 acres are state lands, viz:Clinton, Essex, Franklin, Fulton, Hamilton, Herkimer, Lewis, Oneida, Saratoga, St. Lawrence, Warren, Washington, Delaware, Greene, Ulster and Sullivan

There are three grades of fire wardens. First. The Chief Fire Warden, who is appointed by the Commission. He has "supervision of the town fire wardens, visits and instructs them in their duties, and enforces the law as to fire districts in the towns, and under authority of the Commission commences prosecutions for violation of laws to prevent forest fires. He fills vacancies and removes town fire wardens with the consent of the Commission; has charge of fire wardens' reports, and when a fire is not reported ascertain its origin and result." In addition he has supervision of all bills against the state rendered by the various towns for fighting fires.

Second. The town firewardens are appointed by the Chief Firewarden with the consent of the Commission. Under the Commission a town firewarden is charged with preventing and extinguishing forest fires in his town. In case of fire in, or threatening, forest or woodland the District Firewarden, if any, or, if none, the Town Firewarden shall attend thereto forthwith and use all necessary means to confine and extinguish the same. The Town Firewarden or District Firewarden may summon any resident of his town to assist in putting out fire. Any resident summoned, who is physically able, and refuses to assist shall be liable to a penalty of ten dollars. In case a fire burns over an acre or more of land the fire warden of the town in which it occurs shall forthwith make an examination and report the same on blanks furnished for this purpose to the Commission, giving the area burned over, the quantity of timber, woods, logs, bark or other forest products, and of fences, buildings, and bridges destroyed, with an estimate of the value thereof. He shall also report the cause of the fire and the means used in putting it out.[2] Should the fire

be in his vicinity, although in an adjoining town, it will be his duty to go there immediately and use the same means too extinguish it as though it were in his own town. If no warden from that town be present he shall assume the same authority and the same duties as though the fire was in his own town, until the arrival of the warden from the town within whose limits the fire occurs; but when a warden from the town in which the fire occurs shall arrive he shall assume charge of the fire. No matter which warden orders out the men to fight the fire the town in which it occurs shall be liable for the expense thus incurred. He shall also see that the district wardens are supplied with printed notices, which are furnished by the Commission, containing the rules and regulations relating to the prevention of forest fires, and shall see that the same are posted. He shall be the District Warden of the district in which he lives and a resident of that town. He shall divide the town into districts, if not previously done, and appoint district wardens therein. In dividing the town into districts, mountain ranges, rivers, brooks, and highways are used as division lines. In some cases school districts constitute a fire district. This information is sent promptly to the Chief Firewarden and he locates the district on his map. His pay is $2.50 per day and reasonable expenses, when he works. In accordance with the law he issues permits to farmers to burn fallow and audits the bills against the town for services rendered in fighting fire.

Third. District firewardens are appointed by the Town Firewarden, subject to the approval of the Chief Firewarden. They have jurisdiction over their districts, although they may go beyond when they deem it for the best interest of the forest. They shall see that their district is properly posted with the

---

2.   Section 226—Duties of Firewardens.

cloth fire notices.[3] They do not make a report to the Commission, but assist the town warden in making a report of such fires as may occur in their districts. They shall promptly notify the town warden of any fire in their districts and also report to him a complete list of all men, who may have assisted in putting out a fire, with the number of days or hours each man worked. They receive for their services the same pay as a town warden, viz: $2.50 per day for time rendered. They may issue permits to farmers to burn a fallow, which is situated within their district.

Locomotive sparks and fallow fires cause the larger part of the forest fires. Hence the portion of the law which relates to "Railroads in forest land" and "Fires to clear laud" deserves the most consideration. The section relating to fallow fires reads as follows:

> Section 229: Fires to clear land.—Fallows, stumps, logs, fallen timber, brush or dry grass shall not be burned in the territory hereinafter described from April first to May thirty-first both inclusive, or from September sixteenth to November tenth both inclusive. From June first to September fifteenth both inclusive such fires may be set therein if written permission of the Town Firewarden of the town or District Firewarden of the district in which the fire is set is first obtained. If in a locality near forest or woodland, the fire warden or district firewarden shall be personally present when the fire is started. Such fires shall not be started during a heavy wind or without sufficient help present to control the same, and the same shall be watched by the person setting the fire until put out. Any person violating any provision of this section is guilty of a misdemeanor, and in addition thereto is liable to a penalty of three hundred dollars. This section applies as follows.[4] The section is completed by the enumeration of sixty eight Adirondack and seventeen Catskill towns, which are known as "fallow towns."

The firewarden law has been in force for some time, but not until three years ago, when the office of Chief Firewarden was created, was it effective. Before the Chief Firewarden was appointed there was no one who had the time to look after the large number of wardens; they were a law unto themselves, negligent about reports and some of them lawbreakers. The present law places all the responsibility in a single person—the Chief Firewarden. He in turn holds each Town Warden responsible for reports of all fires and the enforcement of the law, in his town. The town warden in turn depends on his district wardens in the same manner. The law is very weak in that there is no punishment for inability or negligence of firewardens, except their removal. This weakness will exist unless the

---

3. The notices are of white duck 14" by 21" and have the words "Look out for FIRE" at the top in heavy large letters. Below are rules to campers, hunters, fishermen and others in regard to care in building fires, peeling bark, etc. Then follows the law in regard to clearing land. Notices are usually posted on fences, barns, trees and stumps along roads and trails. It often seems like a bit of sarcasm to find one or more of these white cloth fire warnings posted on a charred stub in a burned section, but they are doing a large amount of good.

4. This section was amended by the Legislature of 1903 to read as above. The words " brush or dry grass" were inserted and the time to allow farmers to burn fallow extended ten days in June and fourteen days in September.

office is made a salaried one, when a person would be willing to be responsible for enforcement of the law and attention to his duties, or by placing the responsibility on some town officer, e.g., the Supervisor. It so happens in some towns that the firewarden has so much business of his own that he does not have time to attend to his duties as promptly as he ought. For any inattention, that comes to the notice of the Chief Firewarden, the warden is called upon for an explanation. Frequently wardens are removed or resign and a process of weeding is going on and the force is constantly improving in efficiency. It is always the endeavor to appoint as a warden a man who has not only had experience in fighting fire, but a fellow who can command the respect and obedience of a "posse" of fellow citizens, and has an interest in the preservation of the forest. *The one great trouble is to regulate the use of fire by land owners on their own property.* Many farmers have had no respect or regard for the law and are willing to risk all the woodland they have in order to burn over an acre of fallow. The effect of prosecution has been very beneficial in nearly every case, and it has resulted in letting the careless and lawless know that the time has come when they cannot openly violate the law. However, much has to be done yet as they feel it is an inf infringement upon their right, which they have exercised for generations, when they are restricted in the time that they shall bum. Early in May this year three farmers at Tupper Lake were fined in one day for burning fallow in violation of the law. Another farmer at the same place set a fallow two days later, and after he had been warned. He was also convicted, but this only shows the difficulty and slowness with which the Adirondackers are grasping the new situation. In the case of fallow fires there is the difficulty of proving how the fire originated. Men have been known to hire others to set their fallows afire when they were away and then they return in time to see that all the brush is burned up. To overcome this difficulty an effort was made last winter to have the law amended to read as follows: "if a fire occurs on fallow land in violation of the provision it shall be presumptive evidence that it was set by the owner or possessor thereof." This measure passed the Senate but was over-ruled by the Adirondack members in the Assembly.

The greatest danger from fire is in the spring after the snow leaves until the vegetation is in full leaf, and less so in the autumn after the leaves fall until snow. The intention of the law in prohibiting fallow fires at certain times of the year is to lessen the danger of fire at the two dangerous periods. An extreme case of drought such as has happened this spring in New York State brings up obstacles which any law could not cover as long as there are careless and negligent people at liberty.

The rules on the cloth notices urge the greatest care from hunters, fishermen and others who travel in the woods. As said before fires which burn over more than an acre must be reported promptly to the Chief Firewarden, thus any violation may be taken up and considered while there is still a chance to secure evidence. There has been times in the past when wardens would not report fires or report them as cause unknown in order to shield a guilty party. But this has largely changed and wardens are quite on the alert to fine a guilty party as they receive half of the fine, but not exceeding fifty dollars in any one case.

It is a noticeable fact in the northeastern and middle portion of the state that burned areas are all, or nearly all, in the vicinity of roads, settlements, or farms. The lumberman makes no fires, simply makes everything ready for fires, as they do their work in the winter time. The sooner the residents of the forest realize that the preservation of the forest is their greatest welfare, the sooner will the danger from fallow fires, smudges, camp fires, and carelessness be reduced. Our fires are all caused by human agencies, and if each person in the woods was placed under strict surveillance as to the use of fire then the danger would be largely reduced. The farmers, hunters, fishermen, campers, and last, but not least the railroads burn our woods.

Fire Management Policy in the National Forests—A New Era

1979. *Journal of Forestry*. Vol 77. No. 11. 723–725.

Thomas C. Nelson

Deputy Chief, National Forest System, USDA Forest Service

Abstract: The policy for dealing with fires on National Forest System lands was changed in 1977, in essence from control to management. The change was based on the knowledge and understanding that fire can result in a positive effect on wildland resources. Land managers were directed to provide well-planned fire protection and fire use programs that, in execution, would be cost-effective and responsive to land- and resource-management goals and objectives. After one year of implementation, the net effects of the new policy appear to be positive in terms of both economic and resource values.

The word "fire" is seldom mentioned in laws relating to federal land-management agencies. So, the Forest Service has had to establish its policies from the implied intent of laws containing such words and phrases as *improvement, protection, and securing favorable conditions..*

From the time positive action against forest fires was proclaimed in law by the Organic Administration Act of 1897, the guiding interpretation of the intent relating to wildfire was one of suppression. Each fire was to be put out as quickly as possible. In this way, it was concluded, resource-management objectives would be met and economic loss minimized

That conclusion, probably proper in its day, does not fit the resource-management needs of the 1980s.

Land-management planning, spurred by the Forest and Rangeland Renewable Resources Planning Act of 1974 and the National Forest Management Act of 1976 (NFMA), requires that specific objectives and outputs be established for National Forest System land. This means that deliberate use of fire, as well as control of wildfires, must be an integral part of the planning process and must be responsive to such resource needs as reduction of fuels, improvement of wildlife habitat, and site preparation.

Management must not only meet the ecological needs of the resource and the desires of the public, but it must be accomplished in a cost-effective manner. It must consider the positive as well as the negative effects of both wild and prescription fires on resources. Thus, economic analysis must be an integral part of alternative considerations for fire on any particular piece of land.

As a result of these conclusions, the basic policy on National Forest System lands was changed in 1977.

Historical Perspective

The new policy has its roots in three major developmental phases

The first phase was the creation of a legal base for a suppression organization. This phase included such laws as the Organic Administration Act of 1897, the Forest Fire Emergencies Fund in 1908, the Weeks Act in 1911, and the Clarke-McNary Act in 1924. These laws formed the institutional base for the Forest Service's early control policy, which emphasized rapid initial attack on all fires.

The second phase was assembly of adequate physical plant and manpower. Ironically, the Great Depression of the 1930s was the springboard for development of protection and suppression strategies, along with the appropriate physical plant, to make use of the massive labor force available at the time. Many of these strategies carried over into the 1970s.

The third major phase took place in the years from World War II through the beginning of the space age. Technological innovations—in communications, equipment, and air attack tools—marked this phase. Examples are the National Fire Danger Rating System, behavior models and simulation, and advanced knowledge of fire ecology and effects.

The culmination of the three phases in the 1970s was an effective, technologically sound—and high-cost—fire organization. In 1972, the legal bases, technology, human resources, and finances were combined to achieve a planning objective of controlling all fires at 10 acres or less (Gibson et al. 1976). Coupled with this was the objective to control fires by 10:00 A.M the first day after they started. The basic theory was to achieve the lowest total costs, including resource losses, through an all-out effort to keep every fire as small as possible. This policy did not identify benefits; it merely assumed they existed.

In the 1970s, however, some alarming trends developed. The average number of fires for the period 1970–1975 was 24 percent larger than the average for 1965–1969. The average number of acres burned increased by a similar proportion, and the average expenditures rose 57 percent in real dollars.

Not enough change in climatic conditions occurred between the two periods to explain the difference in fire activity. In addition, it appeared that the amount of fire activity, whether due to increased human use of the forests or other reasons, would have been the same even without the increased spending for costly control systems. In other words, benefits were not matching the inordinate rises in costs.

These trends precipitated a detailed analysis (Gale 1977) that indicated the need for change in the 1980s and beyond. The conclusion was that fire management objectives must be directly related to resource values and the costs of protecting them, and that protection should be commensurate with values and risks. The product should be an accountable, efficient, and cost effective program.

The New Policy

As a result of the analysis, the basic fire policy on National Forest System lands was changed (see Forest Service Manual 5100, 1978). The new policy has three major aims:

• To provide a balanced control program that is cost-effective and commensurate with threats to life and property, public safety risks, and resource output targets.

- To provide for prescription fires, ignited either by plan or naturally, to protect, maintain and enhance national forest resources.
- To provide data, information, and coordination for full integration of fire use and protection in the development, analysis, and evaluation of alternative land-management prescriptions,goals, and objectives

The heart of the new policy is the integration of protection and suppression into the land-management planning process. It should be noted that in some places, particularly the South, the Forest Service had recognized this concept much earlier.

Since land-management planning required by NFMA for all forests is not scheduled for completion until 1983, interim steps have been developed to provide an orderly transition from the former fixed protection objectives to the integrated objectives of the new policy. These steps are designed to assure that the new policy complements, and does not cause conflicts, in areas where land-management planning has been completed. They should also ensure the deliberate and orderly development and refinement of policy implementation.

Implementation

The first step in implementation was the creation of a situation-analysis process for escaped rites. An analysis is made by a multidisciplinary team, and approved by the local land manager, wherever wildfires escape initial attack in areas where land-management or fire-management plans have not yet been established. The analysis is a quick evaluation by local specialists knowledgeable about the resources in the fire area. It evaluates several suppression alternatives on the basis of total cost-effectiveness and the positive, as well as negative, effects of fire on the resources. This analysis and

decision making process is continued through the life of the fire, with one aim being to keep total costs as low as possible. The next step, also implemented in 1978, was to establish fire-management areas and develop interim plans for areas on which land planning is incomplete or inadequate. During the next five years, these interim plans will be superseded by the formal planning for each forest, as required by NFMA. Area plans are developed through a process that emphasizes responsiveness to land- and resource-management goals and objectives; protection from threats to life. property, and public safety; and consideration of cost effectiveness. The process also provides for interdisciplinary and cooperator involvement, and for consistency with the National Environmental Policy Act. With this process, suppression action can be modified or accelerated, depending on what is needed to meet the objectives {USDA Forest Service 1977).

Results

An initial critique, based on the experience of the first fire season with the new policy in operation, indicates positive results. The Intermountain Region of the Forest Service, covering the states of Utah, Nevada, and parts of Wyoming and Idaho, made escaped-fire analyses on 17 wildfires. in comparison with the fixed-suppression objective of the old policy, the new approach showed an estimated saving of more than $800,000 when both suppression costs and net resource damages were considered.

The Maes Creek Fire in the adjacent Rocky Mountain Region offers an example. This fire on the San Isabel National Forest in Colorado began July 5, 1978, in rough terrain above 10,000 feet. After it escaped initial attack, local specialists evaluated potential effects on wildlife, timber, range, aesthetics, recreation, watershed, and soils. From this evaluation, control alternatives shown in table 1 were developed. They ranged from doubling the crew and support strength, as one extreme, to removing all suppression forces and monitoring the area daily from fixed-wing aircraft as the other extreme.

This information and analysis were given to the land manager, i.e., the forest supervisor. He decided the firefighting would be centered around Alternative E, and he set the general objectives. They were: (1) crew safety, (2) control of costs, and (3) adequate suppression efforts to allay the high public concern, especially among the nearby residents of the town of Rye. Acreage burned was not of the highest concern, for the fire was in steep, inaccessible, broken, rocky country and was not advancing rapidly.

The indicated savings over the previous fixed-area and time-period objective was about $500,000. In essence, it was a common sense approach to dealing with the fire cheaply but with adequate attention to the resource objectives and to the public concern and safety.

The other component of the new policy—establishment and advanced planning of fire-management areas—was instituted in all regions in 1978. Regional foresters approved 68 management areas containing more than 4.8 million acres. The control strategies vary from normal deployment of initial attack forces to modified control as limited as simply monitoring fire behavior. The plans analyze the critical site-specific variables in relation to fire-management objectives, and they develop agreed-upon matrices of decisions to guide the strategies in particular situations.

Actual experience with these fire-management areas in adjusting the presuppression and suppression organization resources and finances is limited at this time, because few plans were implemented prior to the 1978 season. But indications are that the net effects, as in the escaped-fire analysis procedure, will be positive (USDA Forest Service 1979).

Table 1. Escaped fire analysis summary, Maes Creek Fire, San Isabel National Forest, July 5, 1978.

	Alternatives					
Remarks	A—Total suppression within 6 days with double present resources (18 crews and support)	B—Total suppression within 8 days with present resources (9 crews and support)	C—Partial suppression within 21 days with 3 crews, 1 helicopter plus support	D—No suppression: monitor status with one crew and 1 helicopter and support	E—No suppression: monitor status with 6 to 8 men and support	F—No suppression: no monitoring except daily air patrol
Estimated control date	7/14/78	7/16/78	7/30/78	8/10/78	8/10/78	8/10/78
Size (acres)	2,300	2,300	2,400	2,400	2,400	2,400
Suppression cost	$694,000	$547,000	$210,000	$225,000	$22,000	$ 2,200
Rehabilitation cost	$ 14,000	$ 14,000	$ 14,500	$ 14,500	$ 14,500	$ 14,500
Estimated total cost	$708,000	$561.000	$224,500	$239,500	$36,500	$16,700

Problems

Of course, a policy change of this magnitude sets the stage for numerous potential problems, and some are developing. They are:

1. A public concern over a less-than-all-out suppression effort in some areas. The reverse is also a problem. It is a misconception that, since wildfires are a natural ecological process, prevention is not important. It must be clearly understood that the new policy is not one of "let burn." Unwanted wildfires must be prevented when possible, and they must be suppressed once they start.

2. The development and implementation of a process for integrating fire management into land management planning is comparatively new, and an unevenness in application is likely.

3. The knowledge and skill to manage fires are not always present. A fire is not being "managed" if it fails to react as anticipated and if objectives are not met. Training and experience, such as that gained from the big fires in Idaho in 1979, as well as the development of additional knowledge of fire behavior, should—overtime—alleviate this problem.

Literature Cited

Gale, R. D. 1977. *Evaluation of fire management activities on national forests.* Policy Anal. Staff Rep. USDA For. Serv., Washington, D.C.

Gibson, H.P., L.F. Hodgin, and J.L. Rich. 1976. *Evaluating national fire planning methods and measuring effectiveness of presuppression expenditures.* USDA For. Serv., Washington, D.C.

USDA Forest Service. 1977. *National fire management policy: final action plan.* Washington, D.C.

USDA Forest Service. 1979. *Proc. National Fire Directors' Meeting:* Portland, Oregon, February 5–9. Washington, D.C.

Is Wildfire Policy in the United States Sustainable?

2007. *Journal of Forestry*. Vol 105. No. 2. 67–72.

Toddi A. Steelman and Caitlin A. Burke

Abstract: Beginning in 2000, wildfire policy in the United States shifted from focusing almost exclusively on suppression to embracing multiple goals, including hazardous fuels reduction, ecosystem restoration, and community assistance. Mutually reinforcing, these policy goals have the potential to result in an ecologically, socially, and economically sustainable wildfire policy that can mitigate the long-term risk of wildfires for human and ecological communities alike. Six years into this new policy, we evaluate the evidence to determine how well the multiple goals are being served. We conclude that suppression and hazardous fuels reduction receive greater attention and resources relative to ecosystem restoration and community assistance. This provides an incomplete solution to mitigating the long-term risk of wildfire, thereby running the risk of perpetuating it.

Keywords: wildfire policy, hazardous fuels reduction, suppression, ecosystem restoration, community assistance

Since 2000, wildfire policy in the United States has undergone significant change. Once driven almost solely by an emphasis on suppression, recent policy has been broadened to include goals for hazardous fuels reduction, ecosystem restoration, and community assistance, in addition to fire suppression and protection. Reducing hazardous fuels before wildfires occur allows managers to put fire back on the land while minimizing risks to people and reestablishing natural fire regimes. Rehabilitating and restoring fire-adapted ecosystems after wildfire occurrence benefits wildlife and habitats, protects watersheds, combats invasive species, and could save millions of dollars in fire suppression costs over the long-term while also minimizing risks to communities, the environment, and firefighters. Promoting community assistance provides support for the industries and workforce that engage in and benefit from ecosystem restoration and hazardous fuels reduction, while also building community capacity to mitigate the wildfire threat. Mutually reinforcing, these policy goals have the potential to reduce the long-term risk of wildfire and result in an ecologically, socially, and economically sustainable wildfire policy. If the building blocks for the new wildfire policy are suppression, hazardous fuels reduction, ecosystem restoration, and community assistance, then the mortar intended to hold them together is collaboration. Congressional direction stipulates that federal agencies, state interests, county officials, and local citizens should work together through collaborative decision making to implement this long-term vision. The rationale is that without communication and collaboration across multiple issues and jurisdictions, implementation would be disjointed and inefficient. Six years after these policy changes were initiated at the national level, we evaluate whether the multiple goals are being met. Our evaluation suggests that although progress has been made, many obstacles remain if the United States is to move toward a more long-term, sustainable wildfire policy.

What Is the Wildfire Problem?

The total amount of forest and rangeland burned each year has risen in recent decades compared with the mid-1900s, and particularly bad fire years are occurring more frequently. More than 2.5 million ha burned each year in 1988, 1996, 2000, 2002, and 2004 (Dombeck et al. 2004). With the increase in acreages burned, more people, property, and infrastructure are at risk.

Natural and social conditions have collided to create the current wildfire problem. In the past 100 years, land-use changes across the country, including fire suppression, logging, road building, and livestock grazing, have led to uncharacteristically high fuel loads, a shift toward severe, stand-replacing wildfires, and an inability of the land to heal itself after disturbance (Dombeck etal. 2004). Recent changes in climate and precipitation exacerbate these trends. Drought conditions ranging from abnormally dry to exceptional drought have occurred throughout large portions of the Intermountain west and southeast for the last 6 years (US Drought Monitor 2006). This is expected to continue as climate change projections forecast warmer temperatures and increased precipitation across North America (Intergovernmental Panel on Climate Change [IPCC] 2001). Even with increased rainfall, higher temperatures boost rates of evaporation and transpiration through plants, increasing the severity of drought and, in turn, the probability, intensity,and severity of wildfire.

Another factor contributing to increased wildfire risk is the growth of human communities living in the wildland-urban interface (WUI)—the place where forests and humans come together. From 1990 to 2000, the greatest increases in houses in the WUI were in the Rocky Mountains and the South (Stewart et al. 2005). Not only are more people and property at risk, but homeowners in these areas expect protection from fire, posing additional challenges and safety risks to firefighters. The final exacerbating factor is what has been termed the "process predicament" or "analysis paralysis" faced by USDA Forest Service decision makers. The Forest Service has asserted that it is severely constrained in its ability to develop and implement projects that could reduce the wildfire threat to communities by the procedural requirements mandated by various environmental laws, such as the National Forest Management Act and National Environmental Policy Act (NEPA; USDA Department of Agriculture Forest Service [2002]). Claims have been made that environmentalists use procedural protocols to delay action (Little 2003). Others claim that Forest Service resistance in addressing legitimate problems with proposed projects results in appeals and holdups (Little 2003). In either case, the excessive time, resources, and effort put toward ensuring the agency is working within its regulatory and administrative framework is preventing the actual on-the-ground work from being completed.

Recent Wildfire Policy Changes

Given the multifaceted nature of the wildfire problem, a policy that focused only on wildfire suppression was inadequate and in need of reform (Busenberg 2004). In the fall of 2000, the Secretaries of the Interior and Agriculture submitted a report to then President William J. Clinton, making recommendations for responding to severe wildfire, reducing the impact of wildfires to communities and the environment,and ensuring sufficient firefighting resources in the future (National Fire Plan [NFP]). These recommendations and the attendant congressional appropriations resulted in an array of technical and financial strategies, guidelines, and projects that became known as the National Fire Plan (NFP), which was, in essence, a $10 billion, 10-year effort to restore forest ecosystems and protect communities. In the 2001 Interior and Related Agencies Appropriations Act (PL 106-291), Congress directed the Secretaries to work with the Western Governors' Association (WGA) to develop a coordinated national 10-Year Comprehensive Strategy for implementing the NFP.

The NFP and the WGA 10-Year Comprehensive Strategy identify four primary goals to reduce the risk of wildfire and build collaboration among all levels of government: (1)improve fire prevention and suppression, (2) reduce hazardous fuels, (3) restore fire adapted ecosystems, and (4) promote community assistance (WGA 2001). The recognition of the need for a comprehensive, integrated approach to wildfire management signifies a critical shift from reactionary policy that focused on wildfire suppression toward a more proactive policy that focused on long-term ecosystem and community health.

In 2002, President George W. Bush announced the Healthy Forests Initiative (HFI),which bolstered the hazardous fuels reduction goal of the NFP by expediting certain projects. The HFI entailed both administrative reforms to reduce procedural delays in implementing fuels reduction projects and legislative action to streamline and prioritize forest health projects. The HFI's central premise was to address the "process predicament" by limiting environmental analysis and the administrative appeals process, thereby allowing the agencies to respond to wildfire risks in a timelier manner (Davis 2004). To that end, the Bush administration issued four administrative changes in December 2002. The first authorized two new "categorical exclusions"—one for high-priority forest health projects and the other for environmental stabilization and rehabilitation projects. This designation eliminates the requirement for lengthy environmental analysis and documentation of project impacts under the NEPA. However, projects must be consistent with the collaborative framework of the 10-Year Comprehensive Strategy, which requires that national, regional, and local authorities and interests work together on technology transfer and decision making to facilitate accomplishments at the local level (WGA2001). The other three administrative actions amended the rules for project appeals to speed review of forest health projects, to facilitate the Endangered Species Act (ESA) of 1973 review of fuels treatment projects, and to conduct pilot tests on the effectiveness of expedited environmental assessments (USDA Forest Service,White House Council on Environmental Quality, and US Department of the Interior 2002).

In November 2003, Congress passed the Healthy Forests Restoration Act (HFRA) (PL 108-148), comprising the legislative portion of the HFI. The HFRA received strong bipartisan support and passed with a 286–140 vote in the House and an 80–14 vote in the Senate, despite several controversial provisions. Title 1, Hazardous Fuels Reduction on Federal Land, encompassed the majority of the controversy and therefore received the bulk of the attention during congressional debate.

The HFRA directs the Forest Service and Bureau of Land Management (BLM) to conduct hazardous fuels reduction projects on up to 20 million ac of federal land, using fire and mechanical methods such as crushing, thinning,and pruning to reduce forest fuels (PL108418). Prescribed fire or wildland fire use (WFU) can be used to reduce fuels and maintain and restore ecosystems. Prescribed fire is ignited by humans, whereas WFU uses naturally ignited wildland fires to accomplish specific resource management objectives. Use of prescribed fire or WFU to reduce fuels is a controversial issue because although it is extremely cost-effective, it also carries significant risk. Under the HFRA, priority areas for fuels reduction projects include the WUI, defined as areas within or adjacent to an at-risk community. The law authorizes $760 million annually for hazardous fuels reduction projects and directs the agencies to spend one-half of what Congress appropriates each year in the WUI. In addition, the agencies are encouraged to focus thinning on small-diameter trees, retaining the larger trees that are most resistant to fire.

A centerpiece of the HFRA is the Community Wildfire Protection Plan (CWPP).To receive funding from the HFRA, a community must have a CWPP. These CWPP planning efforts bring together residents;property owners; local, state, and federal agencies; and others to create and prioritize avision for addressing hazardous fuels treatments in the WUI. CWPPs recognize the importance of community involvement in successful wildfire risk-reduction efforts and allow for more comprehensive collaboration than the NEPA public involvement processes alone. The NEPA collaborative processes typically focus on federal, state, and local agencies and tribes (NEPA Task Force 2003). The intention of the CWPPs is to give communities affected by wildfire a say in public land-management plans and to provide communities a leadership role in identifying the areas for priority hazardous fuels treatment (Newman 2004). The alternatives recommended in the CWPPs are to be adopted by the Forest Service in their NEPA process for subsequent projects.

One of the most debated provisions of the HFRA amended the NEPA to give the agencies a streamlined alternatives analysis procedure (cf. 16 USC 6512 Section 102-108). For each fuels project authorized by the HFRA, the relevant agency must consider the proposed action, a no-action alternative, and one additional action alternative. Where the proposed project is within the WUI, the agency only has to examine the proposed action and one additional action alternative. For projects within the WUI and within 1.5 mi of the community at risk, no alternative analysis is required at all.

Simplification of the administrative appeals and judicial review processes drew criticism from opponents of the Act. Eligibility for appealing a fuels reduction project is limited to those who submit specific written comments on the proposal during the public comment period, perhaps shrinking the pool of contestants, but also establishing consequences for failing to participate early in the decision making process (Davis 2004). Courts are encouraged to expedite judicial review of challenges to hazardous fuels reduction projects, and temporary injunctions pending appeal are limited to 60 days. In reviewing a project, courts must consider the short- and long-term effects of both implementing and not implementing the agency action, thereby favoring action in an at-risk situation.

Together, the regulatory and legislative actions of the HFI/HFRA take a significant step toward freeing the agencies from the analysis and red tape that they say is delaying or prohibiting implementation of hazardous fuels reduction projects. An evaluation of the accomplishments and shortfalls in implementing the four aspects of the NFP, HFI, and HFRA—fire suppression, hazardous fuels reduction, ecosystem restoration, and community assistance—helps us understand whether and how we are moving to a more long-term sustainable wildfire policy.

Evaluating Wildfire Policy Progress

Progress has been made, particularly in the area of planning, funding, and executing hazardous fuels reduction projects. Planning activities are underway to facilitate implementation of hazardous fuels reduction. In 2004 the USFS and BLM issued an interim field guide to help resource managers understand the procedural changes made under the HFI and HFRA (USDA Forest Service 2004). LANDFIRE, a landscape-scale fuels-mapping project that will aid in identifying and prioritizing hazardous fuels reduction projects, has been initiated by the Forest Service and is scheduled for completion in the continental United States in 2008 (US Geological Survey 2006). The HFI administrative reforms are actively being used to expedite hazardous fuel reduction projects. The Forest Service conducted 669 hazardous fuels reduction projects that used the new NEPA categorical exclusion between October 2004 and July 2006 (Bosworth2006). In 2004–2005 the BLM used the HFI categorical exclusions to treat 230,000 ac, and anticipated using categorical exclusions on 1,000 projects covering over 200,000 ac in FY 2006 (Hatfield 2006). Regulations for streamlining Section 7 consultations under the ESA have been finalized; over 830 Forest Service employees have been trained and certified and, through July 2006, over 100 projects have used the new process (Bosworth 2006).

Funding for hazardous fuel reduction has increased, but has not kept pace with the amounts recommended by Congress in the HFRA. For FY 2006, Congress appropriated $907.4 million for implementation of the HFI (this does not include preparedness and suppression spending), including almost $490 million for hazardous fuels reduction, a $25.9 million increase over the previous year's budget (USDA Forest Service 2006). For FY 2007, the Bush administration requested $912.5 million for implementation of the HFI, including $491 million for hazardous fuels reduction (Table 1).

Millions of acres have been treated by the Forest Service and BLM since the initiation of the new wildfire policy. Most of this work has taken place under the HFI, rather than HFRA authorizations. Since 2003 through July 2006, the Forest Service has treated 6 million ac for hazardous fuels reduction and 2.5 million ac for landscape restoration for at total of 8.5 million ac (Bosworth 2006). Of the total acres treated, 65% were in the WUI. In comparison, the number of acres treated using the HFRA Title I authority has been small. In FY 2005 the Forest Service used HFRA authorities to treat 23,000 ac in 71 projects, and in FY 2006 the agency planned to treat 62,000 ac in 138 treatments (Bosworth 2006). Since 2002, the Department of Interior has treated 7 million ac, including 5.9 million ac through hazardous fuels reduction and 1.1 million ac in landscape restoration (Hatfield 2006). Like the Forest Service, the number of acres treated using the HFRA authorities has been limited on BLM lands. In FY 2005, 52 treatments were performed on 9,968 ac and in FY 2006 the agency intended to perform 66 projects on 28,000 ac (Hatfield 2006). For the Forest Service and BLM, most of the HFRA work has taken place using prescribed fire or WFU (Duncan 2006).

The Society of American Foresters and several other organizations developed and

Table 1. Funding for comprehensive wildfire policy activities.

	Comprehensive wildfire policy funding ($000)				
	FY2005 (enacted)	FY2006 enacted)	Percent change FY2006– FY2005	FY2007 (requested)	Percent change FY2007– FY2006
Forest Service					
Suppression	1,043,302	690,186	34%	746,176	8%
Hazardous fuels reduction	262,593	281,793	7%	291,792	4%
Rehabilitation and restoration	12,819	6,188	52%	1,980	68%
Community assistance					
Economic action program	19,032	0	100%	0	0%
Forest health management (federal lands)	14,792	14,780	0%	6,802	54%
State fire assistance	73,099	78,746	8%	56,075	29%
Volunteer fire assistance	13,806	13,683	1%	13,668	0%
BLM					
Suppression	218,445	230,721	6%	257,041	11%
Hazardous fuels reduction	201,409	208,113	3%	199,787	4%
Rehabilitation and restoration	23,939	24,116	1%	24,286	1%
Community assistance					
Rural fire assistance	9,861	9,852	0%	0	100%
Total					
Suppression	1,261,747	920,907	27%	1,003,217	9%
Hazardous fuels reduction	464,002	489,906	6%	491,579	0%
Rehabilitation and restoration	36,758	30,304	18%	26,266	13%
Community assistance	130,590	117,061	10%	76,545	35%

distributed a handbook to help communities prepare CWPPs in compliance with the guidelines of the HFRA. As of March 2006, more than 654 CWPPs covering an estimated 2,700 communities had been completed and approved, with an additional 600 in progress (Bosworth 2006). Stewardship contracting authority has been expanded to allow timber and biomass removal in exchange for services that improve forest health, creating economic incentives for restoration in some places. Since 2003, the Forest Service has awarded 206 stewardship contracts, and the BLM has awarded 114 (Healthy Forests 2006). Two $1 million

biomass utilization projects were initiated in the southeast in FY 2004; in FY 2005 $4.4million in grants were awarded to 20 projects to accelerate the adoption of biomass technologies and create community based biomass enterprises and in FY 2006 18 grants were awarded for a total of $4.2 million (Bosworth 2006, Duncan 2006).

In spite of these gains and the stated shift toward a more integrated, sustainable fire policy, a hierarchy remains that drives fire management activities. Fire suppression continues to be the top priority. The second priority is hazardous fuels reduction, as emphasized in the HFI and HFRA. This emphasis comes at the expense of restoring fire-adapted ecosystems and promoting community assistance.

Identifying Wildfire Policy Shortfalls

Budgets, funding, performance measures, and traditional bureaucratic practices all reinforce the primacy of fire suppression and hazardous fuels reduction over the other wildfire policy goals. A comprehensive analysis of fire funding allocated to Arizona, Colorado, and New Mexico revealed that in 2001 and 2002, the majority of NFP funding in each state went to suppression purposes and hazardous fuels treatments (Steelman et al. 2004). Significantly smaller portions of funding went to ecosystem restoration and community assistance. Moreover, the current agency budget structure and congressional appropriations favor suppression and hazardous fuels reduction. For instance, the ability of the Forest Service to borrow funds from other accounts to cover suppression costs threatens to overwhelm land managers' ability to plan for and address other aspects of the wildfire problem proactively (WGA 2004). Consider, for instance, that in 2002 federal agencies overspent their fire-fighting budget by more than 50%, forcing them to tap other accounts, including those for community assistance, forest thinning, and restoration programs, to pay suppression costs (Reese 2002).

In FY 2007 the Bush administration budget request cut funding for the programs that provide community assistance and assist in ecosystem restoration (Table 1). Although funds for fire suppression and hazardous fuels reduction increased, programs like the Economic Action Program, Forest Health Management, and State Fire Assistance were either eliminated or cut 29–54% from their previous year's allocation (Table1). Rehabilitation and ecosystem restoration funds were cut 68%. Ecosystem restoration

and community assistance programs provide the social and economic capacity to sustain hazardous fuels reduction and ecosystem restoration into the future. Restoring the natural processes and resiliency of forests through thinning, WFU, and other means, although initially a costly undertaking, saves money and effort in the long run by enhancing the ability of ecosystems to recover from natural and human disturbances (Aplet and Wilmer 2003). Offering grants and assistance to communities provides the support to build skills, strategies, and businesses to make these communities self-sustaining. For example, the Economic Action Program has funded marketing research and projects for forest-based micro businesses in communities adjacent to public lands (Sustainable Northwest 2005).

Considering the more than $500 million difference between what Congress authorized annually for hazardous fuels reduction in the HFRA and what has been appropriated each year to date, it is evident that federal coffers can not provide enough funding to reduce all the hazardous fuels that currently threaten communities. In 1999, the Government Accounting Office estimated that it would cost $725 million annually to treat the 39 million most at-risk acres in the United States, using a rather optimistic cost of $300/ac (Gorte 2006). It is not unreasonable for hazardous fuels reduction work to cost $1,000–2,000/ac depending on slope, vegetation, and removal options. Since the 1999 estimates, the number of high-risk acres has climbed to 51 million. If the Forest Service and BLM were to treat all moderate and high-risk acres, the annual cost would climb to $4.3 billion/year, using the same $300/ac treatment cost (Gorte 2006). Without adequate federal funding or the creation of local economic capacity, the Forest Service risks becoming overly dependent on stewardship contracting to fund hazardous fuels treatment projects (Daly

2004). Until feasible markets for low-value forest restoration products are developed, the incentive is for contractors to take larger trees to cover the cost of the work, stressing timber harvest over ecosystem restoration.

Forest Service performance measurements also favor the achievement of hazardous fuels reduction goals over ecosystem restoration and community economic assistance. Tracking "acres treated in hazardous fuels reduction" has become the hallmark of agency success in its new wildfire policy, although clear definitions of ecosystem restoration still have to be identified (WGA 2004). Current performance indicators such as acres treated may not be the best proxy for assessing whether long-term risk is mitigated. Reducing fuels is only one factor in addressing the current wildfire problem. Measuring performance based on tangible outputs diverts managers from performing tasks that are equally important but less concrete, such as building community capacity through economic assistance or establishing partnerships to promote ecosystem restoration (Gregory 2005, DeIaco 2006). Pressure to meet measurable targets results in treating "easy" acres rather than acres that can prevent the greatest damage from being inflicted on communities (DeIaco 2006). A focus on acres treated also may lead to misleading reports. For example, acres that are mechanically treated in one year and then prescribe burned in the next year are counted twice in performance reports (Gregory 2005). Without complete and accurate reporting, land managers, Congress, and the public can not evaluate whether and how resources are being spent to mitigate the long-term risks of wildfire.

The collaborative framework, a centerpiece of implementation of the NFP and the 10-Year Comprehensive Strategy, is not being used consistently at the local, state, and national level (WGA 2004). Collaborative planning was the intended method by which communities could reconcile the multiple, sometimes conflicting, goals for wildfire policy. By spending the time and resources upfront to coordinate information exchange, cooperate on goal setting, and communicate about implementation strategies, agencies could avoid the ill will, appeals, and litigation associated with the process predicament. However, collaborative efforts have been criticized at the state and regional levels as not broadly inclusive, often ignoring those with different interests and objectives, and at the national level as not providing for meaningful participation by nonfederal stakeholders (Daly 2004, WGA 2004, Gregory 2005, DeIaco 2006). Projects from CWPPs, which are meant to help prioritize wildfire mitigation approaches in each locale, are not being implemented on the ground in meaningful numbers (Jensen 2006). At present, there are no data on how many federal land projects identified under a CWPP have resulted in on-the-ground projects, and there is no process to track them (Jensen 2006). Impediments, such as a shortage of financial support and technical resources, lack of national-level recognition of collaborative efforts, in experience of agency employees, and resistance to change within the Forest Service are preventing implementation of the collaborative ideal (Daly 2004, Gregory 2005, DeIaco 2006).

Finally, the progress that has been made in expediting hazardous fuels treatments faces legal challenges. In spring 2006 Chief US District Court Judge Donald Malloy declared the use of the categorical exclusion provision in the HFI/HFRA unlawful. Malloy stated that the exclusions violated the 1992 Forest Service Decision Making and Appeals Reform Act, which requires a public notice and comment process for any national forest project (Scott 2006). In the same ruling, the HFRA predecisional objection process also was invalidated based on an insufficient basis for the public to submit substantive comments on proposed Forest Service projects. The upshot is that some of the procedures for expediting hazardous fuels reduction projects rest on dubious legal ground, thereby jeopardizing their future use.

Policy Implications

By moving away from a strategy based primarily on fire suppression to one that integrates components of fire suppression and prevention, hazardous fuels reduction, restoration and rehabilitation of fire-adapted ecosystems, and community assistance, federal policy seeks to create a more integrated solution to mitigate the long-term risk of wildfire. On paper, the policies put forth in the NFP, the WGA 10-Year Comprehensive Strategy, and the HFI/HFRA are well balanced and address the multiple conditions that contribute to the wildfire problem. But in practice only parts of these policies are being implemented effectively. The focus on suppression and hazardous fuels reduction comes at the expense of the other goals articulated in the NFP and the WGA10-Year Comprehensive Strategy. Without adequate emphasis on the restoration of ecosystems and the development of capacity and incentives through community assistance, it is unclear how sustainable, long-term solutions to the wildfire problem will be feasible.

Hazardous fuels reduction is a necessary but not sufficient solution to the wildfire problem. Without ecosystem restoration, which reestablishes the natural role of fire on the land to result in more sustainable ecological conditions, and community assistance, which establishes local economic foundations for the continued removal of hazardous fuels and ecosystem restoration practices, the federal government and Forest Service will need to provide billions of dollars per year in perpetuity to address the wildfire problem. A sustainable solution must integrate the multiple goals intended in the NFP while giving communities the opportunity to shape them to their specific needs. New policies are not necessary; rather greater attention to the equitable implementation of the existing policy could remedy the current problems. In addition, improved evaluation of the policies and processes will help document whether all the stated goals are served.

What can be done to address these current implementation problems? The first step is recognizing that the NFP, HFI, and HFRA, despite making some important changes, have fallen short of their intended goals. Multiple factors contribute to the current wildfire problem including fire regime disturbance, changes in climate and precipitation, communities expanding the WUI, and gridlock within federal land-management agencies. Fire suppression and hazardous fuels reduction are only pieces of a more comprehensive policy. Greater emphasis must be placed on ecosystem restoration, community assistance, and collaboration to more fully address the complexity of the wildfire issue.

Changes to the institutional infrastructure that support wildfire policy are necessary to achieve the new, diverse goals put forth in the NFP, HFI, and HFRA. These changes need to take place within and among the Bush administration, Congress, and the multiple levels of the public land management agencies. First, the Bush administration, Congress, and the public land management agencies need to treat all goals—suppression, hazardous fuels reduction, ecosystem restoration, and community assistance—as equally important, and show this support through more equitable resource commitments. Second, Congress and the Forest Service should restructure current budget and funding arrangements to support, or at the very least not undermine, the other wildfire policy goals. Creating separate accounts for ecosystem restoration and community assistance that are insulated from suppression spending could help secure funding and foster longer-term, programmatic decision making that is not subject to reductions in bad fire years.

Third, Congress and the public land management agencies should place more equitable emphasis on the measurement of all goals of the wildfire policy. Acres should count after the initial treatment is complete and the land is ready for long-term maintenance burning (DeIaco 2006). Performance measures should link project planning with CWPPs and federal agencies need to provide direction to line workers to prioritize CWPP identified projects (Jensen 2006). Not only will community, state, and environmental interests be served by better accountability, but Congress and the administration will benefit from knowing what is being achieved and what is not being achieved with the money and effort directed to our nation's public lands. Finally, the Bush administration and Congress should allocate more resources to ensure legitimate collaboration as intended by the WGA, especially at the state and local levels, including financial and technical resources and training workshops. Without commitment to the collaborative processes laid out in the policies, wildfire policy will never develop the social infrastructure needed to enable it to be self-sustaining. Tangible evidence that collaboration increases efficiency or reduces the chances of litigation would help make the case that collaboration is worth the time and effort.

Although the Bush administration, Congress, and the public land-management hierarchies are important participants in wildfire policy, they are not the only ones that matter when it comes to implementing policy change. There are numerous examples around the country of efforts to create more sustainable human and natural communities while reducing the risk of catastrophic wildfire (Jakes et al. 2003, Kruger et al. 2003, Steelman and Kunkel 2004). Researchers and practitioners have much to learn from these examples where communities often succeed in spite of existing institutional problems that prohibit a more integrated approach to wildfire management. Documenting these cases, harvesting their experience, and diffusing this information widely can show others how public land-management agencies, state foresters, local fire officials, and various community members can work together to foster place-appropriate change. In conclusion, wildfire policy is at a crossroads. Progress has been made, but more needs to be done to address the multiple causal factors that contribute to the wildfire problem. The current favored alternatives—suppression and hazardous fuels reduction—provide an incomplete solution to the wildfire challenge, thereby running the risk of perpetuating it. Institutional change supplemented by practical lessons from the grassroots can help breathe life into a policy that exists on paper but still has to be given the chance to succeed in practice.

Literature Cited

Aplet, G., and B. Wilmer. 2003. *The wildland fire challenge: Focus on reliable data, community protection, and ecological restoration.* The Wilderness Society, Washington, DC. 44 p.

Bosworth, D. 2006. *Hearing before the Subcommittee on Public Lands and Forests concerning the Healthy Forests Restoration Act implementation.* United States Senate, 109th Cong., July 19, 2006.

Busenberg, G. 2004. Wildfire management in the United States: The evolution of a policy failure. *Rev. Pol. Res.* 21(2):145–156.

Daly, C. 2004. *Hearing before the Subcommittee on Forestry, Conservation* and Rural Revitalization of the Committee on Agriculture, Nutrition, and Forestry. Hearings on the implementation of the Healthy Forests Restoration Act. United States Senate 108th Cong., 2nd sess., June 24, 2004.

Davis, J. B. 2004. The Healthy Forests Initiative: Unhealthy policy choices in forest and fire management. *Environ. Law* 34(4):1209–1245.

Deiaco, R. 2006. *Hearing before the Subcommittee on Public Lands and Forests concerning the Healthy Forests Restoration Act implementation.* United States Senate, 109th Congr., July 19,2006.

Dombeck, M. P., J. E. Williams, and C.A. Wood. 2004. Wildfire policy and public lands: Integrating scientific understanding with social concerns across landscapes. *Conserv.Biol.* 18(4):883– 889.

Duncan, L. A. 2006. *Senate panel seeks details on Healthy Forests Act results.* CQ Green Sheets, July 17, 2006.

Gregory, L. 2005. *Representative of the Wilderness Society testimony before the United States House of Representatives Committee on Resources Subcommittee on Forests and Forest Health. USFS and BLM accomplishments on Healthy Forests Restoration Act.* February 17, 2005.

Gorte, R. W. 2006. *Forest fire/wildfire protection.* CRS report for Congress Order Code RL30755. January 18.

Hatfield, N. R. 2006. *Hearing before the Subcommittee on Public Lands and Forests concerning the Healthy Forests Restoration Act implementation.* United States Senate, 109th Congr., July 19, 2006

Healthy Forests. 2006. *Healthy forests report.* Available online at www.healthy-forests.gov/projects/healthy_forests_report_10_4_06.pdf;last accessed Nov. 28, 2006.

Intergovernmental Panel on Climate Change (IPCC). 2001. *Climate change 2001:Synthesis report.* Cambridge University Press, Cambridge, UK. 408 p.

Jakes, P. J., K. Nelson, E. Lang, M. Monroe, S. Agrawal, L. Kruger, and V. Sturtevant. 2003. A model for improving community preparedness for wildfire. P. 4–9 in *Homeowners, communities, and wildfire: Science findings from the National Fire Plan*, Jakes, P. (compiler). USDA For. Serv. Gen. Tech. Rep. NC-GTR-231. 100 p.

Jensen, J. 2006. *Hearing before the Subcommittee on Public Lands and Forests concerning the Healthy Forests Restoration Act implementation.* United States Senate, 109th Congr., July 19, 2006.

Kruger, L. E., S. Agrawal, M. Monroe, E. Lang, K. Nelson, P. Jakes, V. Sturtevant, S. Mccaffrey, and Y. Everett. 2003. Keys to community preparedness for wildfire. P.10–17 in *Homeowners, communities, and wildfire: Science findings from the National Fire Plan*, Jakes, P. (compiler). USDA For. Serv. Gen. Tech. Rep. NC-GTR-231. 100 p.

Little, J. B. 2003. Un-common ground. *Am. For. Summer* 00:45–51.

National Environmental Protection Act (NEPA) Task Force. 2003. *Modernizing NEPA implementation.* Available online at ceq.eh.doe.gov/ntf/report/final-report.PDF; last accessed September 2003.

National Fire Plan (NFP). 2000. *Managing the impact of wildfires on communities and the environment.* Available online at www.fireplan.gov/reports/8-20-en. PDF; last accessed Mar. 4,2002.

Newman, C. 2004. Community wildfire protection plans from four angles. *J. For.* 102(6):4–7.

Reese, A. 2002. USFS fire accounting questioned. *Land Letter* September 5.

Scott, T. 2006. Judge rejects Bush forest regulations. *Missoulian* April 26.

Steelman, T. A., and G. Kunkel. 2004. Effective community responses to wild-fire threats: Lessons from New Mexico. *Soc. Natur. Resour.*17:679–699.

Steelman, T. A., G. Kunkel, and D. Bell. 2004. Federal and state influences on community responses to wildfire: Arizona, Colorado, and New Mexico. *J. For.* 102(6):21–28.

Stewart, S., V. Radeloff, R. Hammer, J. Fried, S. Holcomb, and J. Mckeefry. 2005. *Mapping the wildland urban interface and projecting its growth to 2030: Summary statistics.* USFS North Central Research Station, Evanston, IL. 38 p.

Sustainable Northwest. 2005. *Investing in rural communities and forest health in the West.* Sustainable Northwest, Portland, OR. 42 p.

USDA Forest Service. 2002. *The process predicament: How statutory, regulatory, and administrative factors affect National Forest management.* USDA For. Ser., Washington, DC. 40 p.

USDA Forest Service. 2004. *The Healthy Forests Initiative and Healthy Forests Restoration Act interim field guide.* Available online at www.fs.fed.us/projects/hfi/field-guide/web/toc.php; last accessed Feb. 2, 2004.

USDA Forest Service. 2006. *FY 2007 budget justification.* USDA For. Ser., Washington, DC. 593 p.

USDA Forest Service, White House Council on Environmental Quality, and US Department of the Interior. 2002. *Administrative actions to implement the President's Healthy Forests Initiative.* Fact Sheet S-0504.02, December11. On file with authors. 4 p.

US Drought Monitor. 2006. *Drought monitor archive.* Available online at drought.unl.edu/dm/monitor.HTML; last accessed Sept. 5, 2006.

US Geological Survey. 2006. *Landfire schedule map.* Available online at www.landfire.gov/schedule_map.php; last accessed June 14, 2006.

Western Governor's Association (WGA). 2001. A collaborative approach for re-ducing wildland fire risks to commu-nities and the environment. Available online at www.westgov.org/wga/ini-tiatives/fire/final_fire_rpt.PDF;last ac-cessed Dec. 21, 2001.

Western Governor's Association (WGA). 2004. *Report to the Western Governors on the implementation of the 10-Year Comprehensive Strategy.* Available online at www.westgov.org/wga/initia-tives/fire/tempe-report04.PDF; lastac-cessed Dec. 16, 2004.

Part 2

Wildfire: Fire Control to Fire Management

■ Elizabeth Reinhardt

The five papers in this part were published in the *Journal of Forestry* or *Forest Science* between 1921 and 1990. Reading through them I was struck by how much the wildfire management profession has changed, yet how much it has stayed the same. There are hints of larger societal trends reflected in these straightforward discussions of wildfire management—for example, only the most recent paper uses gender-neutral terms. I was interested in the vocabulary the authors used and in their implicit but unstated assumptions. and I wondered, where do we go from here? The most recent of these papers is almost 30 years old. How will the next generation of fire managers evaluate our contribution to an evolving profession?

First, we hear from S.B. Show, a "Forest Examiner" with the US Forest Service. Writing in 1921, he presents an analysis of the relationship of slope, aspect, season, and cover type to fire size and occurrence: "Physical Controls of Fires." The paper summarizes data from almost 7,000 fires. Although Show refers repeatedly to "rate of spread," it was not, in fact, a measured or calculated variable. Instead, there is an explicit assumption that bigger fires owed their larger size to the fact that they had spread faster. I have often thought our focus in the fire management community on rate of spread is somewhat misplaced, and I was interested to see it so early in the literature. I was also struck by the fact that Show, in an era before computers and GIS, seems to have had a robust data set.

The second paper, "The Principles of Measuring Forest Fire Danger," from Harry Gisborne, was published in 1936. Gisborne is known as the father of wildfire research, and, in this paper, he introduces the concept of fire danger rating, a concept still in use today, although his particular methods are not. Two things jumped out at me from this paper. The first is, although he doesn't use the term "scientific method," Gisborne calls for a procedure, process, or method that is superior to judgment. Clearly, he has faith in a scientific approach to management. Second is his explanation of why such a process is needed: to increase efficiency, save money, and avoid losses ("too risky exposure of the property"). We are introduced to the idea that there is a tradeoff to be evaluated in fire preparedness, a tradeoff Gisborne characterizes as expenditure versus risk to property.

The third paper, "The Variable Lick Method—An Approach to Greater Efficiency in the Construction of Fire-Control Line," by Robert McIntyre of the National Park Service, was published in 1942. What a wonderful glimpse into WWII-era firefighting it provides! The emphasis on regimentation, regulation, perfect rhythm, and rigid spacing when building line shows a push toward military professionalism that persists to this day. I was particularly struck by the image the author presents of the crew moving in unison, stepping forward with the left foot on the count of one, the right foot on the count of two, and then taking three strokes with their tools, "the tool of each man in unison rises and falls for three working blows on the soil before him." McIntyre adds consideration of safety and efficiency to his discussion.

The military language is even more overt in the fourth paper, "Maintaining an Effective Organization to Control the Occasional Large Fire," written by M.H. Davis in 1953. Amid the military language (the army, the weapons, the general, the buck private), Davis presents a surprisingly modern discussion of preparedness and the fundamentals of an effective organization. The paper discusses the breadth and complexity of the fire control organization, touching on fire weather, training, morale, communications, and leadership.

Finally, "Factors Influencing Forest Service Fire Managers' Risk Behavior," by Cortner et al., written in 1990, is an explicit examination of risk perception and decisionmaking among fire managers. This study looked at factors that impact decisions, where the decisions involve tradeoffs between cost and risk. Risk, while not specifically defined, is presented qualitatively as including "probabilities, values at risk, and costs of avoiding." Risk factors evaluated included safety, resources at risk, policy, public opinion, information reliability, and personal considerations. This paper represents an important departure from the previous papers in several ways: it acknowledges that fire has ecological benefits; it takes a scientific approach to examining human behavior; and it was written during a time when fire policy was more complex and offered more choices and responsibilities to managers than the previous "suppression-era" papers that all implicitly assume that less fire is a better result.

Through all these papers written over 70 years, one sees a desire for professionalism and a faith in science, and these characteristics still define our profession today. Reading early accounts of Forest Service firefighting gives some insight into why this might be. On the one hand was Gifford Pinchot, educated at Yale and in France, an advocate for scientific management. On the other hand were the fire crews: farm boys, "punks, stew bums and pool hall boys" (Tim Egan, *The Big Burn*, quoting Elers Koch). McIntyre notes that "skid row bums and fruit tramps show up poorly." While the forestry profession in the US was grounded in science from the start, the fire management profession was initially more of a seat-of-the-pants endeavor. The scientific methods developed in Europe and adopted by Pinchot regarding managed production of timber had no counterpart in fire management. US fire managers had to look to themselves to develop fire science.

The early papers in this collection implicitly assume that the desired outcome is less fire. As early as Gisborne's 1936 paper we are introduced to the notion that there are trade-offs involved in fire suppression; however, from Gisborne's point of view the tradeoff is simply one of expenditure versus successful property protection. By now we have a much more nuanced view of the tradeoffs involved—tradeoffs between protection of property and ecological values, between public safety and firefighter safety, and, most importantly, tradeoffs between current risks versus future risks. We can acknowledge with Stephen Pyne that "every fire put out is a problem put off," but our behavior does not always reflect this reality.

The adoption of military vocabulary, processes, and culture in fire management is evident in these papers and is still deeply embedded in fire management today. I would argue that, while it has been useful in the past to have a military model for fire management, we could benefit from moving beyond it. When we talk about "fighting fire," we imply that fire is an enemy. While we have come a long way in understanding and communicating the positive ecological benefits of fire, we still portray ourselves as struggling with an adverse force. I hope that when future fire managers look back on our era, they can see a reframing of the role of fire management and fire managers. We can see the first movements toward this in the last paper in this collection, in which the authors describe the possibility of changing managers' behavior. Perhaps instead of seeing our work as people versus fire, we can envision a discussion where fire and people are both elements of a natural system. Fire is a powerful force of nature. When we try to control nature, we set ourselves up for failure and for unintended consequences.

Our role is better seen as tending or managing it and ourselves.

As an analogy, consider the way we might tend or manage spring runoff. Snow accumulates in the high country as fuels accumulate in wildlands. Snowmelt and runoff can damage human infrastructure. They will occur inevitably no matter what our actions. We manage runoff, in some cases, by diverting it to reservoirs and releasing it over time as irrigation water (c.f., deferring wildfire by suppression but reducing fuels through prescribed fires), by enacting zoning and regulation that prohibits building or specifies building standards in vulnerable areas. We wouldn't think of suggesting that we eliminate snowfall or spring snowmelt. We wouldn't think we could divert water to a reservoir indefinitely and never release it—that would be absurd. Our fire-control efforts often show a similar absurdity.

I would like to see us adopt a more pastoral or ecological approach that is not grounded in a vision of success based on putting fires out or keeping fires small, but on restoring fire to landscapes that are currently in fire deficit, and managing human development so that it is less vulnerable to fire. Instead of fire control, we speak of fire management. Perhaps we can go further and substitute the term "tending fire" for "fighting fire." Perhaps instead of touting our 98% initial attack success rate, we can move toward a measure of success that tracks the proportion of fires we did not need to control. The wildland fire management community is moving in this direction, but slowly. The collection of papers here show a profession that has evolved considerably. Today's changing climate, expanding human development in and around wildlands, and the legacy of a century of fire suppression call for continued evolution. The investment in science and desire for professionalism that have served the wildland fire community well for the last 100 years will continue to be assets moving forward.

Physical Controls of Fires

1921. *Journal of Forestry*. Vol 19. No. 8. 917–924.

S. B. Show
Forest Examiner, U.S. Forest Service

Aprevious paper (Notes on Climate and Forest Fires in California)[*] reported on the results of intensive experiments which aimed to determine the effect of certain climatic factors on rate of spread of fires. This report brought out the fact that there was a very specific relationship between rate of spread and wind velocity and moisture content of the litter, and that by experimental methods these relationships could be established.

The present paper is based on work of an entirely different nature. The data used were obtained from the individual fire reports of twelve timber forests in California covering a period of six years, from 1914 to 1919, inclusive. On each individual report, among other things there is given the degree and direction of slope under which the fire occurred. These data were tabulated in connection with an extensive study of forest fires, and an effort has been made to work out the general relation between degree of slope and aspect and rate of spread of fires. The figures derived are therefore general rather than specific in their nature, but it is believed that they represent very closely the relative values which actually exist in general practice. The quality of the data on the individual reports, generally speaking, is good; and it is to be expected that by using very large numbers of individual fires the errors introduced by a few incorrectly reported fires will be largely eliminated. The basis of data is 6,877 fires.

On many of the individual reports the degree of slope is reported as "gentle," "medium," "steep," or "precipitous," using a descriptive term rather than actual degree or percentage of slope. Steepness of slope has therefore been divided into five classes, as follows: level, 0 to 5 per cent; gentle, 5 to 15 per cent; medium, 15 to 30 per cent; steep, 30 to 60 per cent; precipitous, 60 per cent plus.

In tabulating direction of slope, fires on northeast and northwest aspects were grouped with those on north, and similarly fires on southeast, southwest aspects with fires on the south slopes.

As a criterion of rate of spread, the percentage of "C" fires (those over ten acres in extent,) and the size of average fire have been used. The percentage of class "C" fires is an excellent index of differences in rate of

---

[*] Journal of Forestry, December 1919.

spread on the various slopes and aspects. It is easy to see why this is so. On south slopes, for example, with very dry conditions fires will naturally spread more rapidly than on north slopes where the litter is more likely to be moist than on south slopes, and this greater rate of spread will be reflected in the percentage of fires exceeding a given arbitrary limit of ten acres. For size of average fires the same holds true, the size being controlled both by percentage of "C" fires and the average size of "C" fires, which latter varies in much the same way with slope and aspect that other criteria and rate of spread do.

Relation of Rate of Spread to Aspect and Slope

Table No. 1 shows percentage of class "C" fires on slopes of different degree and of different aspect. It is to be noted that the percentage of "C" fires increases quite rapidly as the slopes become steeper, the graphic relation approaching a straight line in form. The values on level land are the lowest, as is to be expected. The differences between north and south slopes are particularly striking, while east and west slopes keep an intermediate position and in most cases are fairly close together.

The percentage of "B" fires (one-fourth to 10 acres) is practically constant for all slopes and aspects, averaging 35 per cent (range 33 to 37 per cent) so that percentage of "A" fires (0 to one-fourth acre) is a reciprocal of percentage of "C" fires.

It is to be noted that on the average there are 38 per cent class "C" fires on south slopes as against 21 per cent on north slopes and 11 per cent on level land.

Table 2, based on the same fires as previously used, shows size of average fire on degree and direction of slope. As in the case of percentage of "C" fires, the aspects rank: Level, North, East, West, South. For each aspect the average fire increases directly with percentage of slope, the rates being:

Increase in acres per 10 per cent of slope— North, 16 acres; South, 60 acres; East, 33 acres; West, 45 acres.

It is interesting to note that of the total acreage burned (Table 2) 56 per cent is due to south slope fires, only 6 per cent to those on level land, while north, east, and west slopes are about equal.

The relative average sizes of all fires on the various aspects vary greatly, south slops having the largest and level the lowest.

In Table 2 is also given the average size of all fires on degree of slope alone. Roughly, those on gentle slopes are 2½ times as large as on level land; on medium slopes nearly twice as great as on gentle; on steep, twice as great as on medium. The data for precipitous lopes are very fragmentary and unsatisfactory, only a few fires having been recorded as on such situations.

If the data in Table 2 be platted on degree of slope corresponding to the percentages used, and curved, it will be found that the curves have a slight upward trend, in the same ratio that percentage of slope per degree increases with increase in steepness of slope.

The relations so far discussed are, obviously, very general, and if the data were available it would certainly be profitable to express rate of spread on the basis of forest type instead of aspect. That unquestionably is the manner in which rate of spread will finally be worked out, but at present our data on occurrence of fires are not sufficiently well tied in to timber type to permit of such a division. All that this study claims is an expression of the general relations between aspect and degree of slope and rate of spread.

It is clear enough that on a north slope at low elevation, yellow pine type, rate of spread will be greater than on a south slope, high elevation, in red fir type. At present, however, with no further apologies, the data are presented for what they may be worth.

Table 1. Percentage of C Fires on Degree of Slope and Aspect

Aspect	Slope								Average per cent	Total number
	Gentle		Medium		Steep		Precipitous			
	Per cent	Total number fires	Per cent	Total number fires	Per cent	Total number fires	Per cent	Total number fires		
North	19.0	303	19.0	808	30.0	235	26.0	23	21.0	1,369
South	25.6	476	36.0	1,335	56.8	379	62.2	37	37.8	2,227
East	17.4	218	29.4	514	48.4	126	33.3	9	29.0	867
West	18.8	170	29.3	453	54.7	106	20.0	5	30.5	734
Level	..	..	..	..	..	..	..	..	11.2	1,680
Average	21.4	1,167	29.5	3,110	47.9	846	46.0	74	26.1	6,877

Table 2. Size of Average Fire on Degree of Slope and Aspect (Size in Acres)

Aspect	Slope										Per cent of total acres
	Gentle	Total acres	Medium	Total acres	Steep	Total acres	Precipitous	Total acres	Average	Total acres	
North	34	10,250	62	49,605	83	19,590	29	660	59	80,105	12
South	85	40,320	146	194,975	327	124,115	292	10,825	166	370,235	56
East	55	11,995	83	42,605	191	24,045	31	280	91	78,925	12
West	57	9,680	123	55,750	223	23,625	170	850	122	89,905	14
Level	..	..	..	..	..	..	..	..	25	41,500	6
Weighted Average...	62	72,245	110	342,935	226	191,375	171	12,615	96	660,670	100
Simple Average...	58	..	103	..	206	..	131	..	..	..	...

Work at present under way will establish relative rates of spread in the different timber types in the State.

Seasonable Differences in Occurrence of Fires

Table 3 shows the relative seasonal importance of fires on different aspects. In deriving these figures the total number of fires in a given month is taken as 100 per cent, and the percentage occurring on each aspect was compiled with this as a basis. It is seen that the figures for south slopes are high in May and June, low in July and August and rise again through September, October and November. The figures for east and west slopes are, generally speaking, practically constant throughout the season, while those for the level land show a general tendency to be high in the early part of the season and low at the end. Relative percentage on north slopes rises from May to August and then drops to end of season.

A study of one o f the charts made shows what of course everybody knows, that in the early and late part of the fire season a very high percentage of the fires are on the warmer south slopes and that during the peak of the season the north and south slopes are more nearly equal than at any other time.

Another expression of this same relationship is given on another chart. There are, in the data used, twelve forests for six years, or a total of seventy-two points for each month. If, for example, on every forest and every year fires occurred on the south slopes in August, regardless of the actual number of fires, the value would be 100 percent of the possible. Likewise, if only twenty-four points were represented the relative value would be 24 divided by 72 or 3 per cent. The values used on this chart are obtained in this way: It is seen that, beginning with May, fires occur over twice as often on south slopes as on north slopes, that through June and July the two approach more and more closely, and that in August they are practically the same; then, during September, October, and November the lines again diverge quite sharply, showing that more and more the fires tend to occur on south slopes. On the average fires occur only 80 per cent as many times on north slopes as they do on south.

A still further and very striking difference between north and south slopes is found in the total number of fires which have occurred during the six-year period. On the north slopes a total of 1,369 fires have been reported and on the south slopes 2,227 fires, or about 62 per cent more. It maybe that there is a difference of 10 or 15 per cent in the area of land on north and south slopes in the National

Table 3. Relative Seasonal Importance of Fires on Different Aspects

Month	North	South	East	West	Level	Total	Per cent of total fires
May	14	41	11	6	28	100	3
June	19	35	12	11	23	100	11
July	21	29	12	11	27	100	24
August	23	29	12	11	25	100	34
September	19	35	12	10	24	100	17
October	16	40	14	11	19	100	9
November	11	48	13	11	17	100	2
Per cent of total fires	20	32	13	11	24	100	100

Class of fire	Per cent in brush	Per cent in timber
A	22	47
B	36	35
C	42	18

Forests, but we can only conclude that a difference in number of fires as great as 62 per cent is a significant difference. This is due, not only to the fact that fires do not occur so often on north as on south slopes, but probably also to the fact that a considerable number of fires burn out of themselves on the north slopes while this is a comparatively rare occurrence on the south slopes. At any rate the wide divergence in the values for the two is too great to be explained entirely on the grounds of coincidence.

In this connection it is interesting to refer back to the paper cited earlier. It was found that fires can not spread if moisture content of litter is over about 8 per cent, and that one season on typical north and south slopes the moisture content was above danger point one-fifth of the time on south slopes and nearly one-half of the time on north slopes. In other words the two entirely independent lines of investigation check in at least a qualitative measure.

Relation of Rate of Spread to Cover Conditions

One exceedingly important control of rate of spread has not yet been discussed; namely, the influence of cover conditions. In the analysis of the fire statistics all fires

occurring for three years were segregated into two classes, those occurring in the timber and those in brush. Among the latter were included some fires which were essentially brush fires, although strictly speaking they occurred in timber. In so far as suppression is concerned, a fire in dense underbrush in an open timber stand should be classed with brush fires rather than timber fires and this practice has been followed. As can be seen from the following table, very striking differences in rate of spread were found for the two sets of conditions.

Under both sets of conditions the percentage of B fires was the same, but on the average there is over twice as high a percentage of C fires in brush as in timber and, conversely, the percentage of A fires is twice as high in timber as in brush.

Years ago Supervisor Wynne worked out comparative rates of spread for fires in brush and in timber and found that, on the average, brush fires on slopes up to 40 per cent and wind velocity up to 8 miles per hour spread 51 acres per hour, while timber fires under the same range of conditions spread 24 acres, or something less than half as much. These two entirely independent lines of investigation give a ratio of about

Table 4. Percentage of Possible Number of Times Fire Could Occur (Basis, 12 Forests—6 years)

Month	North slope	South slope	Difference
May	20	43	23
June	65	82	17
July	86	96	10
August	96	97	1
September	82	92	10
October	43	60	17
November	10	28	18
Average	40	50	10

two to one as representing differences between brush and timber fires.

It will be seen at once that one very important reason for the high percentage of class C fires on south slopes is in the greater percentage of brush on such slopes, as compared with north slopes. It has been impracticable to tabulate the data to a point where percentage of C fires in timber and brush on the various aspects can be determined, but there can be no question that character of cover as well as climatic conditions makes the south slope the most dangerous with which we have to deal.

It seems reasonable to suppose that the cause for the occurrence of brushfields on south slopes to a greater degree than on others lies in the fact that weather conditions are more severe there than elsewhere. Once brushfields are established, they themselves tend to continue the difficulty of control of fires and perhaps accentuate the differences between north and south slopes.

Further work, not only on the relative spread of timber and brush fires but of fires in all the cover types, will be necessary to elaborate and complete the study here reported on.

Summary

The study of the influence of degree and direction of slope and of type of cover on rate of spread of fires in California shows:

1. That on south slopes the per cent of class C fires is about twice as great as on north slopes and three and one-half times as great as on level land.

2. That east and west slopes throughout the State occupy an intermediate position between north and south slopes.

3. That size of average C fires and of all fires varies in the same order and about the same degree that percentage of C fires does.

4. That percentage of C fires varies directly with per cent of slope, as does size of average fire.

5. That in the early and late parts of the fire season a very high percentage of all fires occurs on south slopes, while during the peak of the season north and south slopes are about equally represented.

6. That covering a period of years 62 per cent more fires have occurred on south slopes than on north slopes.

7. That on the average fires occur on north slopes only 80 per cent as often as on south.

8. That fires in brush spread more rapidly than those in timber and that this is at least partially responsible for the higher percentage of class C fires on south slopes, on account of greater proportion of brushfields on such sites.

The Principles of Measuring Forest Fire Danger

1936. *Journal of Forestry*. Vol 34. No. 8. 786–793.

H. T. Gisborne
Senior silviculturist, in charge, Division of Forest Management, Northern
Rocky Mountain Forest and Range Experiment Station, Missoula, Mont.

The current adjustment of fire control action to the prevailing status of fire danger is basic to
efficient forest fire control. Unless this is done, expenses will be too high in some years and
losses will be unnecessarily great in others. It is therefore essential that the prevailing status
of fire danger be determined accurately and with as great refinement as is warranted by the
flexibility of the fire control organization. The article describes the technique of applying the
device known as a fire danger meter to obtain a numerical rating of fire risk.

Research in fire danger measurement was commenced in 1922 at the Northern Rocky Mountain Forest and Range Experiment Station of the U.S. Forest Service, with headquarters at Missoula, Mont. Since then investigations have been made concerning (1) what to measure, (2) how to measure, and (3) field use of these measurements. In all cases the laboratory restricted field experiments have been followed by several years of extensive application and test on the 10 fire forests of northern Idaho and western Montana, comprising an area of some 17,000,000 acres.

From this work three basic principles have become evident. They are:

1. All the significant daily variables of fire danger must be measured or dependably estimated.

2. These measurements must thoroughly sample the forest property.

3. The net effect of the several variables must be determined by some method so that, whoever applies it to the measurement for a certain day and area, the same rating of danger will be arrived at

These principles not demand that certain factors be measured in all forest types that a certain number of stations be used per unit of area, or that use of experienced judgment be eliminated. The research and field tests

Fire on the Land   **51**

have shown, however that unless all the significant factors are considered, unless each fire and climatic type is properly represented, and unless the applications of personal judgment and estimate are standardized, the resultant ratings of fire danger will not be as accurate or refined as they can be made by adherence to these principles.

Factors That Affect Fire Danger

Studies of going fires and the analysis of fire reports have shown for the Northern Rocky Mountain Region that at least five factors affect the fire danger, which varies from day to day and season to season.[1] These variable factors are: (1) date, or season, (2) fuel moisture, or inflammability of specific materials, (3) wind, (4) visibility range, and (5) activity of fire-starting agencies. Some of these factors include two or more sub-factors, but each of the five listed is of major significance in this region

Date, or Season.—Even though temperature, humidity, wind, and fuel moisture may be the same in mid-June as in mid-July or even mid-August the green vegetation such as grass, weeds, and brush is maturing, curing, and becoming less a fire retardant and more a fire accelerator as the season progresses. Hence danger increases with date, up to a certain point. An allowance is made for this variation in the Northern Rocky Mountain scheme. Even more consistent with calendar date is the number of hours of dangerous burning weather, according to hours of sunshine each day.

On September 15, for example, there are at Missoula, Mont., 3.3 fewer hours of sunshine and 3.3 more hours of cool, calm, humid weather favorable for fire control than there are on June 15. This shortening of the fire day opposes and finally off sets the effect of maturing vegetation. Both the vegetative factor and the hours-of-sunshine factor are, at present, brought into the rating according to calendar date.[2]

Fuel Moisture.—This variable, which determines forest inflammability, is the second most important one considered by the Northern Rocky Mountain method. The top layer of duff and half-inch-diameter dead branchwood are the two fuels measured. The drier these fuels, the greater the danger, and in determining current danger it does not matter whether this dryness is controlled by precipitation alone, humidity alone, or any combination of precipitation, temperature, humidity, wind, and sunshine. Fire research in this region has shown (1) that if surface duff or litter moistures and the moistures of small dead branchwood (2) are measured, the inflammability of a majority of the dead fuels is accounted for. A statistical analysis (4) of the influence of weather factors on the moisture content of these fuels has shown that even very complete weather measurements cannot be used dependably every day for this purpose.

Some of the finer fuels such as tree moss, dead grass, and weeds also contribute appreciably to fire danger when they are extremely dry. As the moisture contents of these fuels change faster than duff or twigs and lag only a few hours behind relative humidity, the Northern Rocky Mountain method provides a higher rating of danger whenever the humidity drops below 15 per cent.

Wind.—Many experienced men believe that wind is one of the most important variables of fire danger. Cases can be cited of crown fires occurring with snow the ground, and of blow-ups during high humidity. In all such cases a high wind is usually the cause and when the fuels are dry and the humidity

---

1. Topography and fuel type (3) are factors of fire danger which vary from forest to forest.
2. The fact is recognized that vegetation starts growth earlier and stays green and is a fire retardent much later in some years than in others. A research project is being carried on in an attempt to determine more accurate vegetative criteria (5).

low, even a small increase in wind velocity immediately accelerates the rate of spread of fire. Wind velocity, therefore, cannot be omitted from any complete scheme of rating fire danger. In the Northern Rocky Mountain scheme wind is given almost as much weight as fuel moisture.

Visibility Range.—The distance at which small smokes may be detected is a factor in fire danger. Visibility conditions may be such as to permit seeing small smokes 20 miles or more from a lookout,or the atmosphere may be so hazy that new fires can occur within 1 mile of a lookout, yet not be seen. When visibility is restricted there is greater danger of fires becoming large, more lookout stations must be manned, and in dry weather more men must be sent to every fire that escapes quick detection In the early spring, before the average season fire control organization is warranted, the only action needed may be the placement of a few observers at their stations, their distribution depending primarily upon atmospheric visibility range.

Fire-Starting Agencies.—Fire danger and fire control are, of course, affected by the activity of any fire-starting agency. Lightning, which causes about 72 percent of the fires in Region One, is therefore brought into the Northern Rocky Mountain scheme.

The fire records show that man does not produce peak loads of fires in Region One. Consequently this scheme does not rate danger higher on week-ends and holidays even though there are more people in the forests at such times. An allowance is made, however, for increased danger whenever numerous land-clearing fires, permitted by law, are occurring adjacent to the forest property being protected.

These five factors, date, fuel moisture, wind, visibility, and certain fire-starting agencies have therefore been selected as the principal variable controls of fire danger in Region One. On any midsummer day only one of these may contribute to fire danger, while the four others subtract from it. Or any two factors may contribute while three subtract.

In fact, if we distinguish merely between two classes, the favorable to high danger and those favorable to low fire danger, without any regard for the degree of favoritism, 120 combinations of these five factors are possible. Hence it is not surprising that forest managers have in the past failed to agree when, for instance, date and duff or wood moisture have been conducive to high fire danger while humidity, wind, lightning, and visibility were favorable to easy forest protection.

Classes of Fire Danger

With five factors combining their effects or counteracting each other, the integration cannot be left to personal opinion or judgment. This is especially true if large expenditures of funds are to be based on the resultant opinion, or if the daily opinions are to be combined at the end of a fire season into a seasonal rating to be used in determining efficiency of fire control. To rate fire danger some method must be used, such as Koppen's system of classifying climates,so that whoever applies the system will arrive at the same result as another using the same basic data. The more numerous the factors, the more essentials such a system become.

Each significant degree or class of danger must then be designated in such a manner that all danger reports are consistent and comparable. There is little consistency and less comparability in designations such as "none, easy, average, bad, and very bad" when applied by numerous men of different temperament and varying observational experience.

The Northern Rocky Mountain scheme divides the total range of danger into seven classes. This number was chosen because in this region there are at present seven rather definite stages or steps in fire control organization. These areas follows:

Class 1. No men need be specially detailed to fire control.

Class 2. Fire control stations covering heavy slash or active brush-disposal operations should be manned. A few key lookouts also may be manned following lightning storms early in the season.

Class 3. All key detection stations should be-manned and the limited number of guards and smokechasers included in the "minimum protective organization" should be sent to their stations.

Class 4. The "average season protective organization" should be placed.

Class 5. The "first overload" positions should be filled.

Class 6. The "second overload" positions should be filled.

Class 7. Every economically justifiable step should be taken, including the mobilization of supplemental overhead, stationing crews of firefighters at strategic points, and other action specified by the Region One fire-control plan for meeting the most extreme fire danger.

Actually there is no sharp division between one class of danger and another, and in actual practice the fire control organization is not expanded suddenly from one stage to the next. Perhaps 100 gradations could be set up, but for all practical purposes the present distinction of seven classes of danger and seven stages of organization are as refined as is needed at present

In addition to the advantage of correlating fire control action with degree of danger, the numerical scale of one to seven serves another very practical purpose. Previous to the adoption of this scale forest managers had no standard terminology whatever for describing fire danger. Usually, the more expletives used, the greater the danger, but when expletives had to be eliminated the reporting officer often had difficulty in specifying extreme and critical conditions.

The provision of a definite scale was immediately recognized by field men as an aid in describing the status of fire danger consistently and understandably. It was a material factor in the substitution of fire-danger measurements for fire-danger "guesstimates" on the National Forest in Region One.

Fire Danger Meter Designed

The integration of effects of the five outstanding factors of fire danger so that all of the 120 combinations can be expressed on a scale of 1 to 7 requires the use of some sort of homograph, alignment chart, or slide rule. A simple device, called the Harvey exposure meter, has long been used for a similar purpose in photography, and this idea was used in designing a fire danger meter, illustrated in Figure 1.

This pocket-size cardboard device is easily adjusted to register the status of each .of the six factors of fire danger. The resultant class of danger is then indicated on the scale of 1 to 7. Although this device is complex at first glance, its manipulation is not difficult after a few minutes' practice. It has been readily adopted in Region One.

For best daily and seasonal use a record should be kept to show for each day the status of each factor of fire danger and the resultant class of danger. Figure 2 shows the type of chart used in Region One for this purpose during 1933, 1934, and 1935. The values plotted on this chart are the 1935 daily averages for all the stations used in this region. An improved chart has been designed for use in 1936.

With a record of this kind available for a ranger district, a National Forest, or an entire Region, the forest manager is aware, during the season, of the existing status and the current trend of each element of danger. The meter then shows the fire control organization warranted or justified.

On the 10 fire forests of Region One, the measurements of factors are made in early May at less than a dozen stations. These are

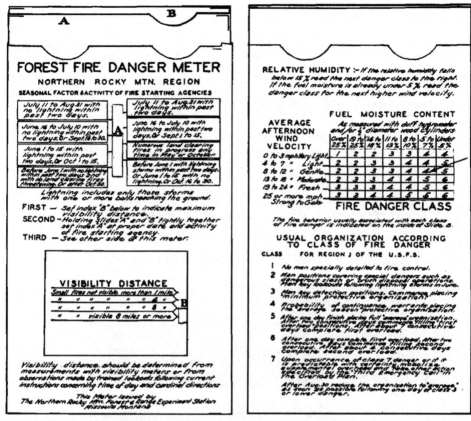

Figure 1. The forest fire danger meter.
Left: front view. At the top are two slides, marked A and B. These slides are movable and are read through windows in the outside of the holder.
Right: back view. The dimensions of the folder are 3 7/8 inches by 6 3/4 inches.

at the lower elevations. By early July or after the lookouts have been placed, as many as 180 lookout, smokechaser, and ranger stations measure one or more elements of fire danger. Approximately 70 stations serve as the major network for the 17 million acres included in these 10 Forests. This represents an average of one complete weather and inflammability station per 240,000 acres. During the past three fire seasons this distribution has appeared .to be sufficient to reveal nearly all local differences in danger warranting different fire control action.

At the close of each fire season the daily record of fire danger may be used to rate the character of any part of the season,

or of the season as a whole. As shown by Figure 2, in July and August, 1935, there were 0 days of Class 1 danger, 2 days of Class 2, 10 days of Class 3, 22 days of Class 4, 27 days of Class 5, 1 day of Class 6, and 0 days of Class 7. Multiplying the number of days by its class value gives a total of 263, which divided by the total number of days, 62,gives an average danger of 4.24 for this period.

Other records indicate that the rating for the easiest probable July and August would be about 2.8, while that for the worst probable would be 5.5. Class 2.8, therefore, represents zero danger and 5.5 represents 100 per cent for this period. On this scale, the

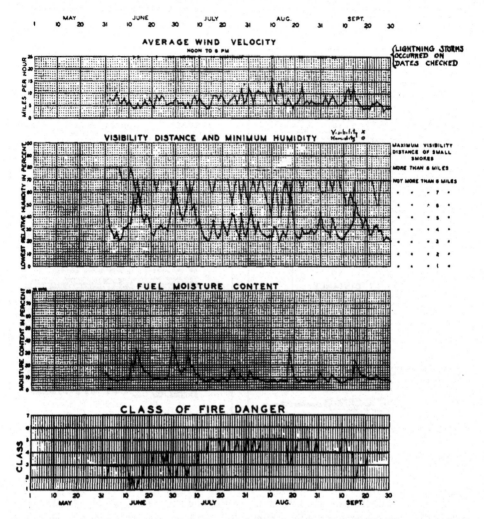

Figure 2. Type of chart used to record fire danger in ten National Forests in Region One during 1935.

1935 average of 4.24 represents 53 per cent of worst probable danger.

This method, therefore, makes it possible to express the character of fire danger on a percentage basis which is readily understood by boards of directors, state legislators, and congressmen who are not familiar with, and perhaps not interested in, the detailed technique of danger measurement. Rating on a percentages scale also permits ready comparisons of successive seasons on a Forest, or comparisons between Forests for a particular season.

For example, July and August rated as 46 per cent of worst probable in 1933, 71 per cent in 1934, and 53 per cent in 1935. The individual National Forests could be likewise compared.

In comparing one year with another, we need to consider length of fire season as well as intensity, because a four month season rating, say, 50 per cent of worst probable may justify greater organization costs than a two-month season rating 65 or 70 per cent. The records for Region One do not, however, as yet include a sufficient variety

Table 1. Efficiency of fire control

Total costs plus damage	Character of season — Per cent of worst probable									
	10 or less	11 to 20	21 to 30	31 to 40	41 to 50	51 to 60	61 to 70	71 to 80	81 to 90	91 to 100
	Efficiency A = Acceptable. E = Excellent									
Dollars										
250,000	A	A	E	E						
315,000		A	A	E	E					
400,000			A	A	E	E			Too risky	
500,000				A	A	E	E			
630,000	Too costly				A	A	E	E		
800,000						A	A	E	E	
1,000,000							A	A	E	E
1,260,000								A	A	E
1,600,000									A	A
2,000,000										A

1. These relationships might apply on a forest area of 25 to 30 million acres requiring fire control facilities of a type and quantity necessary in a region of fire behavior similar to that in Region One.

of seasons to indicate the best method of incorporating the length-of-season factor.

Rating Fire Control Efficiency

As an indication of the use of seasonal ratings to determine efficiency of fire control, Table 1 shows several relationships that are sometimes overlooked.

First is the fact that even during the most favorable fire seasons some minimum $250,000 assumed in this case must be expended for the maintenance of fire control facilities and a skeleton fire control force of men.

Second, as the character of the fire season becomes more dangerous the control force must be increased proportionately, or the property will be unduly exposed to damage.

Third, the ultimate objective of efficient fire control is to follow the very narrow path between acceptable expenditures and too risky exposure of the property.

Fourth, as danger increases more leeway should be allowed to compensate for those unfavorable catastrophes that so often characterize critical seasons. In Table 1, a 91 to 100 per cent season allows a leeway of $1,000,000 in total costs plus damage,

while a 31 to 40 percent season restricts efficient work to a range of only $250,000. The bare minimum of maintenance costs is acceptable, with no allowance for damage, if the character of season rates as less than 10 per cent of worst probable.

Fifth, such a scheme definitely specifies the maximum total cost plus damage that will be accepted as efficient for each specific class of season.

Without some such system of comparing cost plus loss with character of fire danger, efficiency ratings can be at best only impartial and thoroughly experienced judgments. If the judge has incomplete information concerning the character of season, if he is unfamiliar with fire behavior and fire-control practices in the region in question, if he is a personal friend or a personal enemy of the fire-control manager being judged, or if the judge's health happens .to be below .par at the time, his rating of efficiency of the work is not likely to be as just and impartial as it would be if the criteria of efficiency were deliberately specified.

By measuring fire danger and by testing the relationship of cost-plus-loss to character of season, it should be possible to produce experience tables similar to Table 1 which will permit the rating of fire control efficiency by the same principle that life and fire insurance companies use in judging a human or property risk. These companies could not operate for long if they let each of their many agents judge each risk on the basis of the agent's limited and very personal experience. Such judgment and rating should be a procedure inherent in the business, not a personal ability of the employee.

By the methods described here it is possible to determine the prevailing status of fire danger far more accurately than it can be estimated, and on this basis to adjust the size of the fire control organization so that unjustifiable expenses and unnecessary losses may be reduced. Later it may be possible to rate efficiency of fire control much more precisely than is possible by personal judgment alone.

Literature Cited

1. Gisborne H. T. 1928. Measuring forest fire danger in northern Idaho. U.S.D.A. Misc. pub. No. 29, 63 pp.

2. ——. 1933. The wood cylinder method of measuring forest inflammability. Jour. For. 31: 673–679.

3. Hornby, L. G. 1935. Fuel type mapping in Region One. Jour. For. 33: 67–71.

4. Jemison, G. M. 1935. Influence of weather factors on the moisture content of light fuels in forests of the northern Rocky Mountains. Jour. Agr. Res. 51 (10): 885–906.

5. ——. 1936. The effect of low vegetation on the rate of spread of fire in the northern Rocky Mountain region. Manuscript. Thesis prepared under a Charles Lathtop Pack Fellowship at Yale University.

The Variable Lick Method

An Approach to Greater Efficiency in the Construction of Fire-Control Line

1942. *Journal of Forestry*. Vol 40. No. 8. 609–614.

Robert N. McIntyre
National Park Service

Much attention has been devoted in recent years to increasing the efficiency of crews engaged in the construction of fire-control lines. The "variable lick method" described in this article is an outgrowth of experiments undertaken in 1936 with a 40-man C.C.C. crew on the Snoqualmie National Forest. It has subsequently been tested with fire-fighting crews of different sizes in both eastern and western Washington and should be of particular interest to forest-protection agencies in connection with the present emergency.

Description of Method

The "Variable Lick Method" is a progressive method of managing men through regimentation to obtain a maximum output of control line per man engaged. Its essential features are as follows:

1. An entire crew of control-line construction men move forward along a fire line without changing their relative positions in line.

2. A fixed or predetermined percentage of the available man power is delegated to do a particular job in line construction and is equipped for that task. For example, the rule of thumb for the Pacific Northwest is to have 20 per cent of the crew made up of axmen, 20 percent Pulaski-tool men, and 60 per cent hazel-hoe men in the eastside pine type; with 25 per cent axmen, 25 per cent Pulaski-tool men, and 50 percent hazel-hoe men in the westside Douglas-fir type.

3. The spacing of workmen within each unit is a variable which must be kept above a minimum for working safety and below a maximum for the sake of efficiency. The spacing limit for axmen is from 8 to 12 feet and for hazel-hoe men from 4 to 6 feet. Trained axmen

hold a spacing of 10 feet very well, and trained hazel-hoe men space well at 4 feet.

4. The clearing crew of axmen work as individuals in a progressive line, but not by the individual assignment method. The crew is led by an *advance line locater* who blazes a route ahead of it. He is kept informed on rate of fire spread by one or more scouts. Next comes the working crew headed by the *ax-line locater* who marks trees, logs, and other material that must be cleared from the line. He is continually on the move and is backed by his line of axmen. No single axman handles all of a major task; in his continual forward movement he begins, adds to, or finishes one or more of the jobs encountered in clearing. The last man on the clearing crew is the *pacer*, who regulates the speed of the whole crew by oral orders to "speed up" or "slow down," which are passed up the line man by man. When the crew becomes swamped with work in a thicket or when the digging crew catches up with the axmen, he calls for help from the reserve of Pulaski-tool men.

5. The Pulaski-tool crew is the emergency unit which functions in both clearing and digging as the need arises. This crew is headed by a *Pulaski-line locator* who leads his men up to aid the hard pressed axmen over a tough stretch of cover and then, when their services are no longer needed, orders them to resume line digging behind the axmen. In the process of digging line his crew functions in the same manner, order, and procedure as explained for the larger digging crew in the following paragraph; with the exception that it should begin digging on a count of "seven" to allow the larger crew an opportunity to catch up and thus form one digging crew.

6. The digging crew of hazel-hoe men headed by the *line locater* work as a unit, step by step, blow by blow, regimented by a count, spoken aloud by picking individuals, which causes a foot to be placed ahead along the line or a working blow to be struck by each tool. For example, a hazel-hoe crew of 20 men, spaced at intervals of 4 feet, face the fire and will build control line to their right. Count "five" has been called for all to hear by the *line locater*. The *pacer* begins his count, "one," and the left foot of each individual is placed ahead to their right along the fire line. On count "two" each man lifts his right foot and places it in such a position as to give him working stance. Counts "three, four, five," are now called slowly, and the tool of each man in unison rises and falls for three working blows on the soil before him. Counts "one, two," ring out again, and feet are placed forward into position; on counts "three, four, five," three more blows are struck by each man in the crew.

In a few moments the rhythm picks up speed and the line of diggers is building control line by a semi-painless process. Concentration on placing the feet and cutting sod by the "count method" keeps the realization of muscular fatigue in the background.

The *line specifier* may find that the quality of control line exceeds specifications; if so, he signals the *pacer* orally for a "count four." The count is now changed and the men, keeping in mind that counts "one, two" are walking counts, proceed with line building at a faster rate because one working count has been eliminated.

If tougher going is encountered several hundred feet ahead, the *line locater* calls for a count of "six." Now for this stretch of line the crew moves ahead more slowly than before because of the addition of two working counts. The "Variable Lick Method" thus makes possible the building of a control line to certain specifications that can be controlled by the *line specifier* who inspects the work of and follows behind the digging crew.

When a digging crew is fairly well trained by the count method of lick regulation the *pacer* can be eliminated. Count "four" or "six" or "ten" can be called by either the *line locater* or *line specifier*, and with a little help from picked individuals on "one, two" for walkings steps, the working blows will be struck in perfect rhythm.

Crew Organization

The following tabulations indicate the equipment and responsibilities for each individual in a 10-man and in a 20-man crew.

The basic organization of a 40-man crew will be practically the same as of a 20-man crew. Two foremen, however, are needed. Another crew must follow up for bucking out, firing out, holding backfire, progressive mop up, and patrol. All listed positions on the crew become more specialized. Men assigned to the ax crew, Pulaski-crew,and hazel-hoe crew should be about double those listed for the 20-man crew. Where a count of eight is needed for a 20-man crew to cut and dig a 12-inch line to mineral soil, a 40-man crew needs a count of three or four.

Advantages

1. The method has been well tried in the field over a period of 5 fire seasons. It can be used efficiently with a group of from 8 to 48 men.
2. Control line to mineral soil can be built to specifications and maintained for the working period over different types of ground cover.
3. The most work with the least effort is obtained from the digging crew through strict regimentation by use of the "counting system."
4. Every man on a crew, including the foreman, is a worker contributing his bit to the finished product.
5. The thrill of forging ahead into each new task, as well as being able to look back and see the great accomplishment resulting from cooperative effort, builds a high morale and zest for the obstacles ahead.
6. The accident rate per man engaged is very low due to rigid spacing. When accidents do occur they can be chalked up against faulty training in use of tools, infrequent rest periods, or dull tools.
7. Danger of the crew being trapped by the fire is reduced to a minimum by using an experienced straw boss as *advance line locater*. He is in a position to warn the foreman of imminent danger.

Organization of 10-Man Crew

Number of men	Title	Fire line job	Tools carried	Responsibility
1	Foreman	Foremanship	2½-pound D.B. ax, first aid kit C.S., 1-quart canteen, 4 fusees, Bantam shovel when ax is carried on belt.	Make decisions, call rest periods, act as auxiliary pacer for digging crew, iron out kinks, keep up morale, do all burning out if no crew follows.
1	Straw boss	Advance line locating	2½-pound D.B. ax, 1-quart canteen, whetstone, 5-gallon water bag (soaked), and tin cup.	Scout fire line and blaze ahead for ax crew, stay close to fire, remember line to be burned out behind, help axmen, get water to crew.
1	Axman	Clearing	2½ or 3½-pound D.B. ax, 8-inch file, 1-quart canteen.	Chop out logs, trees, brush to 6-foot width, but throw out nothing. Call on Pulaski-man for help when necessary. Keep close to fire.

The Variable Lick Method

Number of men	Title	Fire line job	Tools carried	Responsibility
1	Pulaski-tool man	Clearing	5-pound Pulaski-tool, 8-inch file, 1-quart canteen.	Throw out all brush and logs, aid axman when called, loosen and dig out all chunks, clip off all small brush, do not crowd man ahead.
1	Pulaski-tool man	Clearing and digging	5-pound Pulaski-tool, 8-inch file, 1-quart canteen, pocket first aid kit.	Finish all work left by the man ahead, and if none help on digging crew. Bark and dig hole under any logs crossed and not chopped out. Watch for low limbs, trim them off.
1	Hazel-hoe man	Line locating, pacing, and digging	Hazel-hoe (8-inch blade), 1-quart canteen, 8-inch file, supply of salt tablets.	Spell off line specifier on counting, change count as needed for cover type, always dig by pulling tool down hill. Keep men spaced to 4-foot minimum. Take short steps on counts of "One" and "Two."
4	Hazel-hoe man	Digging	Hazel-hoe (8-inch blade), 1-quart canteen.	Dig line to mineral soil, 12-inch tread, make each blow count, strike no rocks, rake them out, never lift tool above head. "One" and "Two" counts are walking steps, keep in rhythm, watch man ahead for 4-foot spacing.
1	Clean-up man	Line specifying, pacing, and line clean up	Fire rake, Kortick, or McLeod tool, 1-quart canteen, pocket first aid kit, and 8-inch file.	Spell off line locator on count, when clean up is much call higher count and vice versa. Find count that will build line to specifications given by foreman. Stop crew to help him hold back fire.

Organization of 20-Man Crew

1	Foreman	Foremanship	2½-pound ax, first aid kit C.S., 1-quart canteen, 4 fusees, Bantam shovel.	Make decisions, call rest periods, act as auxiliary pacer for digging crew, iron out kinks, keep up morale, burn out if no crew follows up, stop crew to hold line.
1	Straw boss	Advance line locating	2½-pound ax, 1-quart canteen, whetstone, 5-gallon water bag (soaked), and tin cup.	Scout out fire line and blaze ahead for ax crew, stay close to fire, remember line is to be burned out, help out axmen, get water back to crew as soon as possible.
1	Axman	Ax line locating	2½ or 3½-pound D.B. ax, 1-quart canteen, pocket first aid kit, and 8-inch file.	Mark all material to come out of trail 6 feet wide, limb trees not to be cut, keep axmen spaced at 10-foot intervals, move ahead slowly until word comes to speed up. Help axmen.
1	Axman	Axman clearing	2½ or 3½-pound D.B. ax, 1-quart canteen, and 8-inch file.	Chop out logs, trees, brush to 6-foot width, but throw out nothing. Watch for marked material. Do not crowd.
1	Axman	Pacing and clearing, clean up	2½ or 3½-pound D.B. ax, 1-quart canteen, and 8-inch file.	Throw out all material cut ahead of you, finish clearing job if you can, if not call on Pulaski crew to help. Speed up or slow down axmen in accordance with work left for you or Pulaski men to do.

Number of men	Title	Fire line job	Tools carried	Responsi
1	Pulaski-tool man	Pulaski line locating	5-pound Pulaski-tool, 1-quart canteen, 8-inch file, and pocket first aid kit.	Keep your men dig the hazel-hoe men help the axmen. Th throw out everythin line until your crev Start digging line left off helping ax of "seven" to allov to catch up and one single digging
2	Pulaski-tool man	Clearing or digging	5-pound Pulaski-tool, 1-quart canteen, and 8-inch file.	Dig in rhythm wi crew until your li you up to aid the ay 10-foot spacing thro and dig out all sur and trees left in a sume digging on tl longer needed to l Space at 4-foot inte count "seven" unt crew catches up forms a continuous
1	Pulaski-tool man	Pacing, clearing, digging, and clean up	5-pound Pulaski-tool, 1-quart canteen, and 8-inch file.	Help count as pacer is digging line alor hoe crew, act as cle they are digging alc axmen in the job o
1	Straw boss	Line locating, and pacing, digging	Hazel-hoe (8-inch blade), 1-quart canteen, 8-inch file, and pocket first aid kit.	Spell off line speci change count as n type, always dig down hill. Keep n 4-foot minimum. T on counts of "one"
9	Hazel-hoe men	Digging	Hazel-hoe (8-inch blade), 1-quart canteen, and 8-inch file.	Dig line to miner tread, make each b no rocks, rake then tool above head. and "two" are counts, keep in rh man ahead for 4-fo
(Two of these men can be used for buck-ing out and falling behind clean-up man if no crew follows.)				
1	Hazel-hoe clean-up man	Clean up and digging	Hazel-hoe, fire rake, Kortick, Mc-Leod tool, or Bantam shovel, 1-quart canteen.	Clean out all narr line and test questi mineral soil, keep on digging crew. count. Line specifi over a tough spot
1	Straw boss	Line specifying and pacing	Bantam shovel, first aid kit (P.S.), 1-quart canteen, 5-gallon	Spell off line loca when clean up is toc

8. Time taken to teach a digging crew to function properly is reduced to a minimum by using the "counting system." The author has lined up, equipped, and given a digging crew of local woodsmen satisfactory practice run within 15 minutes from the time that they reported for work. An experienced foreman can put either a digging crew of local woodsmen a working crew of C. C. C.'s through initial training and have them producing satisfactory control line within 30 minutes. Initial training for local axmen in an emergency can be accomplished within 30 minutes if the foreman works with them.

9. A trained crew can outbuild and outlast any crew of similar size using either the "individual-assignment method" or the "man-passing-man method." Quality of line produced by crews of equal numbers of men rules out the "one lick method" as a competitor, although the total amount of line built is comparable when our crew is using a count of "three" or "four."

10. Large amounts of specified control line can be expected of a trained crew of from 8 to 48 men, providing that they are backed up by a crew of equal size which will buck out, burn out, hold, mop up, and patrol the line they build. A 40-man to 48-man crew with training can be expected to build control line, with 12 to 16 inches of mineral soil showing at a rate of 1 mile per hour for the first 2-hour work period in a medium heavy pine type. A trained 20-mancrew can make the same line at 2,000 feet per hour for the first 2-hour period in the same type. A 10-man crew is less efficient but will outperform and outlast any other crew having the same number of men working as individuals

11. Tools as outlined in this article have given the greatest amount of efficiency in actual performance in a variety of ground and timber types. Fire rakes, Council or Rich tools, Kortick tools, and McLeod tools have been used by the different digging crews with moderate success. Most crews using them have changed to the hazel-hoe (8-inch blade) for straight digging and have retained the others for clean-up work. Hazel-hoes fitted with straight tapered handles are not satisfactory; the light, slim recurved handle is the more efficient even though more of them are broken during a working season. D.B. axes of 2½ pounds have become the special tool for axmen in the Pacific Northwest (toughest of timber types). Trained men can accomplish more work with this ax in the variety of ground cover, brush to 6-inch trees, and logs removed, than with any other tool used. Crews not followed by buckers or fallers should use 3½ -pound D.B. axes for the heavier work.

Disadvantages

1. Since it is probably the most technical of the methods using a progressive line of workingmen for building control line, it should not and cannot successfully be used by in experienced foremen or fire bosses. Only trained foremen can teach inexperienced crews in its proper use. It is better not to use the method at all than to misuse it.

2. Although local woodsmen and C. C. C. crews adapt themselves well to the method, skid-road bums and fruit tramps show up poorly. After the first hour of work the latter get the idea that their production of control line is above the rate of pay received. "Goldbricking" and reigned illness soon cause the crew to be undermanned and inefficient.

3. An inexperienced crew will soon become leg and tool weary. For the sake of safety, work or training should not be prolonged at any onetime. New crews cannot be trained for more than 2 hours at a time if their morale is to be kept at a peak. Even an experienced crew should be given rest periods or a "fiver" every 2 5 minutes. Only in an emergency should a trained crew be expected to produce control line for more than 4 hours at any one stretch.

4. The method can create a serious physical hazard to members of the crew. The excitement and heat of the fire line, competitive urge to speed ahead, lack of physical stamina, and the excessive drinking of water without taking salt tablets to keep up a salt balance in the blood stream have c used crew members to collapse and become a serious liability.

5. The method as outlined is only for the initial clearing and digging crew in the progressive method of fire control. An equal number of trained men under competent foremen should follow up the "variable lick crew" and perform the tasks of bucking out large logs not chopped out, felling leaning snags, burning out the line, knocking down hot spots , holding and widening the burned-out line, mopping up that line for a short distance into the fire, and finally patrolling all sectors until the general mop-up crew arrives.

6. The method as outlined is not applicable to the direct method of line building, as a rule. Other methods of line building, such as the "two foot" on slow-burning fires, the "parallel" on hotflashy fires, and the "indirect" on crown or reburn fires, fit the method well. The use of this method therefore implies that at least 80 per cent of the control line must be burned out to insure its use.

Maintaining an Effective Organization to Control the Occasional Large Fire

1954. *Journal of Forestry*. 1954. Vol 52. No. 10. 750–755.

M. H. Davis
Assistant regional forester, Southwestern Region,
U.S. Forest Service, Albuquerque, New Mexico

Those of us who have been exposed to fire protection through the years are well aware of the hazards of gambling with preparedness to handle fires quickly and effectively, when they occur. Fire is no respect or of precedent—geographically, topographically, nor with respect to day of the week or month of the year. If there is a theme to this paper it consists of one word *preparedness*.

Small fires normally can be handled effectively by one or a few men equipped with shovels, axes, backpumps, and the basic tools. The fire which makes for trouble, however, is the one which finds us unprepared.

An effective organization for any program is competent and prepared for action. The objective, here, is to define preparedness for the forester, landowner, or fire boss responsible for handling the occasional large fire.

How can we assure preparedness? First of all it is essential to assume certain controlling facts, i.e.:

1. The area involved constitutes a potentially hazardous fire area at intervals.
2. The area is under protective control of a designated individual, organization, or agency.
3. The facilities essential to fire suppression are available and authority is vested to take action.
4. Financial responsibility for suppression costs is fixed.

The party or agency responsible for action had better determine a date—appropriate to the area concerned, and with due regard to fire burning periods—by which time the fire "team" is ready to go. Recurrent, annual, thorough analysis of factors is imperative to the maintenance of an effective fire organization. There is no substitute for this essential.

Fire Organization

What are the factors? They include:

1. *Man-power.*—Top to bottom—"general" to "buck" private. Are they available and ready to go?
2. *Weapons.*—An army without guns and ammunition is hardly an effective force. Let's not fight fire with antiquated equipment. Development, testing, and adoption of modern fire-fighting equipment is one of the best assurances of an effective suppression job.
3. *Transportation.*—The "army" and the "weapons" won't do any good if you can't put them on the fire in time.
4. *Food and supplies.*—Don't ever try to fight a fire on an empty stomach.
5. *Communication.*—Have you been in the back-country—50 miles or so—and in urgent need for reinforcements, a doctor, aerial reconnaissance,or just plain "contact" when your radio or telephone facilities failed?
6. *Information.*—Call it intelligence, scouting, reconnaissance, or what you will. A fire boss on a big fire is lost unless this information is reliable, current, and complete. Let's be sure to give high priority to fire weather forecasts and interpretation in terms of fire behavior.
7. *Dispatching.*—I have never forgotten an experience during my early years as a dispatcher in southern California. A modest 'fire of several hundred acres was burning in the back country and the ranger was on the job. I heard not a word from him on the fire itself for three days, from the time he left his station. Final]y, he reported: "Fire under control—see you soon." When he came out to civilization I asked why he hadn't kept us informed. He replied "I

knew what you would do, and you did it much obliged." He was referring to the fact that the "sinews of war", men, supplies, equipment, etc., were forthcoming. What he failed to recognize was that the job had to be done the hard way. The moral to this story is that dispatching should be recognized as a vital factor in the fire suppression job. To be most effective, it requires team work, i.e., by the dispatcher and by the fire boss, or his designated contact, with the dispatcher.

8. *Liaison.*—Sightseeing traffic in the vicinity of a fire can be a real problem. Two years ago it was necessary to secure a CAA ban on unauthorized airplane traffic in southern New Mexico, due to smoke caused poor visibility conditions in the vicinity of the fire suppression operating area. With four aircraft working dawn to dusk dropping equipment and supplies over a 40,000-acre area, danger to life and aircraft was very real. How about public relations—the press, radio, police, and persons "in-the-path"and the near-by property owner who wants to back-fire to save his property—also, the logging companies and the military? These all are liaison jobs.
9. *Safety.*—Last, but perhaps of greatest import, is the safety of the men on the job. Too many tragic examples of injury and death exist to permit secondary consideration of this item.

The foregoing factors are merely elements of the job. What about on-the-line organization, over-all strategy, and tactics on the job? I suspect that the big fires become bigger because of one of the following reasons:

1. Poor generalship—strategy, leadership, organizational ability—or let's call it competency.

2. Failure in tactical operations.
3. Failure in organization—too little, too late, or inefficiently utilized?
4. Failure in logistics—dispatching, transportation, or equipment.
5. Failure in training or preparedness.

I have not listed the above in priority order. However, I am confident that you will agree that failure in any one or more of these factors or elements can, and all too frequently does, result in the"blow-up" of a fire situation. The elements to remember are:

THE BOSS

Organization
Strategy } Direct on the line use of the "sinews of war"
Tactics

THE "SINEWS"

Manpower
"Weapons"—equipment, tools.
Transportation
Food and supplies
Communication
Information
Dispatching
Liaison
Safety

Fighting a large fire is a big job. It is complex. It is a science involving a knowledge of fire behavior, the elements, organization, psychology of relationship, the integration and direction of a complex organization of specialized effort and the well-being and safety of people.

An effective organization must be maintained to handle efficiently any large fire. The most competent man available should be placed in command of the fire suppression job. This is a fundamental essential.

In directing on-the-line use of the "sinews of war" this man must base his decisions involving organizations, strategy, and tactics, upon factual information acquired personally or for him by delegated assignment. We are familiar with most of these factors—topography, cover, slope, fire location, etc. I want to place emphasis on another item of great importance which too often is passed over lightly, i.e. fire weather forecasts and interpretation.

Wide variation exists in availability and use of fire weather forecasts. For example, California utilizes mobile fire weather forecasting units and secures on-the-ground-factual data and forecasts. These data can be scientifically applied to fire behavior patterns and used in determining strategy and organization.

Arizona and New Mexico, however, are dependent upon generalized weather forecasts which lack sufficient localized data upon which to evaluate fire conditions, as in the case of California. Emphasis on adequate fire weather forecasts and adequate interpretation and their effects on fire behavior provide the fire boss with essential tools in determining strategy and organization needs. Tactical practices like wise will be effected. Moreover, safety of men can be assured to a far greater extent by the aid of adequate fire weather forecasting and interpretation in relation to fire behavior.

Another factor of extreme importance on the project fire is the adequacy of inspection. Many fires have been "lost" because of a weak "link-in-the-chain." Had the weak link been discovered early and remedial action taken promptly would it have become "the big fire of that year?" Effective training, supervision, morale or team-work, and all other vital elements are wasted if the fire boss is not sure that strategy, tactics, organization, communication, etc., are being executed as planned—and successfully. The fire boss must know that there are no weak spots. On a major fire authority and responsibility must be delegated. There are competent and incompetent inspectors. Selection, training, assignment, reporting standards, and on-the-spot effectiveness are vital requirements.

The elements of the fire suppression job must be understood and listed prior to the time of need. Their application, analysis, and solution must be written and a fire plan for the area developed. This fire plan should set forth all details of organization including source of manpower. It should provide all essential data for all members of the "team"—from the fire boss on down—to know their assignment, to whom they are responsible, and specifically what their duties will be on the fire. The experience and background of all key members of the organization should be listed and training needs set forth.

A word about the tenure of employment for key men—effectiveness of fire suppressions hampered too often by failure to retain experienced, trained, and competent men. Security of employment is a prerequisite to retention. Smokechasers, crew leaders, and technicians acquire competency only through continued training and experience. Such men should receive full consideration in assignment to other work.

A training program for all key personnel should be prepared and revised annually.

This program should detail each year (a) objectives, (b) methods, (c) time and place, and (d) participants. Safety for men in fire suppression should not be ignored. Instructors should be assigned sufficiently in advance to assure adequate preparation and. if possible,advance review of training by the instructors with the training officer should be arranged. Instruction should cover, in addition to basic elements, the following: (1) escape routes for emergency "exit"; (2) lookouts for safety warning in cases where danger may exist; (3) avoidance of critical spots during inactive or rest periods; (4) routes of travel into the assigned position and also emergency exit from the assigned position; (5) safety precautions involving use of tools. The factor of safety is a must in suppression tactics. The leader, foreman, sector, division or fire bosses, machine operators, and specialists must be personally competent and assume responsibility and authority to prevent accidents.

An excellent training device is to detail trainees to going fires on a programmed, planned basis. This was done in 1951 in southern New Mexico where a group of trainees from three regions were pooled under assignment to the regional training officer. Training consisted of orientation and job assignment. Orientation included briefing on a 40,000-acre fire including organization, strategy, tactics, aerial operations (smokejumpers, supplies), safety, inspection, etc. Assignment was subsequently made to specific line or staff jobs in accordance with needs of the trainees. As apprentices, the trainees were detailed to instructors. As opportunity afforded and progress warranted final assignment was made. Results of this program for the 64 trainees were most encouraging.

It is recommended that a review be made of each year's fire record and fire suppression accomplishment, to determine strong and weak spots in the program. There should be also an annual fire report of the year's effort. Such a report will show trends, accomplishments, problems, and other data essential to complete understanding of the fire protection situation, for the area concerned.

As professional foresters we should not accept any standard short of professional adequacy for any phase of our professional activities. This is particularly significant with respect to fire suppression. As an aid to fire suppression preparedness the following checklist is presented.

1. What is the fire history for the area?
2. What is the fire record for the past five years?
3. What are the necessary equipment re requirements?
4. Is the fire plan for the ensuing year adequate and up-to-date? Does it cover current hazard and risk?
5. Is the fire boss competent?

6. Does the fire plan provide adequately for the elements outlined above? A—1 to 3 (the boss)and B—1 to 9 (the sinews)?

7. Have training plans been prepared for the current year and instructors and trainees assigned?

8. Have the "Board of Directors" for the area been briefed adequately as to financial requirements to maintain a competent "ready" suppression force?

9. Can the fire boss put his X and initials on a statement to his "Board of Directors" that "We are prepared ?"

Two final important items often overlooked are morale and physical readiness. Morale is a potent force. It consists of far more than pay, housing, food and the 1-3's of employment. Good leadership is essential to effective fire suppression. Can the fire boss pass this test? Physical condition of a fire crew and the key men in the organization can be assured through proper planning and work direction. These factors should not be left to chance. So, (1) analyze, (2) define objectives, (3) plan, (4) train, (5) inspect, (6) direct, and remember safety, morale, and physical condition.

Factors Influencing Forest Service Fire Managers' Risk Behavior

1990. *Forest Science*. Vol 36. No. 3. 531–548.

Hanna J. Cortner, Jonathan G. Taylor, Edwin H. Carpenter, and David A. Cleaves

Abstract: Fire managers from five western regions of the USDA Forest Service were surveyed to determine which decision factors most strongly influenced their fire-risk behavior. Three fire-decision contexts were tested: Escaped Wildfire, Prescribed Burning, and Long-Range Fire Budget Planning Managers first responded to scenarios constructed for each decision-making context. Various types of risk were manipulated in each context to determine what factors could influence a shift in risk behavior. Following the presentation of scenarios, managers rated and ranked decision factors that might influence their decision-making on fire. Results show that safety, the resources at risk, public opinion,and the reliability of information were important influences on manager decisions. Local or regional policy changes and personal consideration had less influence. Manager ratings and ranking of what factors are important in fire decision-making are consistent with fire-risk decisions taken in each of the three decision contexts. Fire-risk behavior also varied from one geographic region to another and from one fire-decision context to another. Depending on the kinds of risks managers perceived, their decisions shifted along the risk-avoidance/risk-taking continuum For. Sci. 36(3):531-548.

Keywords: Fire management, forest fire, fire policy, decision-making.

Until the 1970s, Federal Land Management Agencies responded to the risks presented by wildland fire hazards with an explicit risk stance—suppress fires as quickly as possible. For example, the suppression activity standard adopted by the USDA Forest Service in the mid-1930s—and used by most public land management agencies—called for control of every fire on the day it was reported or by 10 A.M. the following day. This "10 A.M." suppression policy was supplemented in 1972 by a policy under which presuppression activities would be designed and selected to restrict fires to 10 acres or less in size program performance was evaluated on the criterion of the number of fires kept under the 10-ac minimum. Adopting the assumption that all fires were bad, both policies minimized risk and became part of the justification for increasing protection forces.

Adopting this risk-averse posture in fire planning resulted in escalating costs—a 57% real increase in presuppression expenditures for the Forest Service from 1970 through 1975 (Gale 1977). In the same period—characterized by a series of severe fire years—the number of fires and acres burned also increased. The proportion of fires below the 10-ac target size changed little. It became apparent to Congress and to the Office of Management and Budget that aggressive employment of presuppression resources did not necessarily avert fire-induced damages and suppression costs—at least not to the extent assumed in the 10-ac planning criterion. Reducing the proportion of fires less than 10 ac in size by 2% would have taken an estimated 90% increase in presuppression expenditures (Gale 1977).

Guided by research which demonstrated the necessity and benefits of fire in natural ecosystems, and prompted by concerns over the costs of sustaining risk-averse fire suppression, the federal and management agencies to change their fire management policies. The National Park Service began to modify its policies on fire suppression and prescribed burning in 1968. The Forest Service followed suit in 1978—changing from a policy of fire control to one of fire management (Nelson 1979, Kilgore 1984). Since that time policy has evolved to cover human-caused lightning caused, and planned ignitions in wilderness, parks, and national forests. In these new fire policies, the risk stance is no longer explicit. Managers are presumed to be more sensitive to the tradeoffs among resource values, the risks associated with the fire hazard, and the costs of fire protection and suppression activities.

The policy debate generated by the 1988 fire season raised the specter that the policy innovations of the past 2 decades would themselves be at risk. The 1988 fire season burned near record acreage; suppression efforts cost over one-half billion dollars (Fire Management Policy Review Team 1988). Fires in the Yellowstone are dramatically characterized the severity of the fire season. These fires, with their images of walls of flame endangering nearby towns and Old Faithful, received nightly television coverage. There were public and congressional criticisms that agency policies allowing some naturally occurring forest fires to run their course might have run amok. Had Yellowstone's values as a scenic and recreational attraction been destroyed by failure to implement swift and aggressive suppression? Or would the following spring show that the fires were actually less destructive than initially estimated and that indeed there would be long-run beneficial impacts (Lewin 1988, Shear 1988, Bolgiano1989)?

In response to the public controversy the Secretaries of Agriculture and Interior established an interagency fire policy review team to assess the fire management policies of the public land management agencies and to recommend changes needed to address the highlighted problems

The Fire Management policy Review Team's report, issued on December 14, 1988, concluded that basically the agencies fire management policies were sound. But they also noted that those policies needed to be refined, strengthened, and reaffirmed (Fire Management Policy Review Team 1988). Included among the specific suggestions were recommendations for a comprehensive set of criteria for deciding whether or not to allow natural ignitions to burn as prescribed fire, a clear description of the process to ensure adequate public involvement and intergovernmental coordination, and the improved raining of fire program personnel. The issues identified as requiring further analysis included determination of the adequacy and consistency of fire cost analysis and risk assessment procedures. The review team also recommended that no prescribed natural fires be allowed until fire management plans were revised to meet the recommended criteria (Wakimoto 1989).

Implementation of the findings of this— or any other—policy review can be facilitated when those trying to effect behavioral change have a clear understanding of the factors that motivate those whose behavior they are trying to change. This suggests the utility of understanding the relationship between fire managers' perceptions of the values at risk in fire management situations and how fire managers actually make risky decisions. Do fire managers exhibit tendencies to be risk-taking or risk-avoiding various fire management situations? What decision factors weigh most heavily upon those decisions: fire managers' perceptions of safety, the resources at risk, public opinion policy, the reliability of information, or personal considerations? What decision factors are likely to effect changes in the managers' risk behavior, i .e., moving from a risk-taking posture to a more risk-averse posture and vice versa? Using a survey of Forest Service fire managers, this study quantitatively addresses these questions. Answering these questions can assist decision-makers in identifying the factors that they must appeal to in order to influence risk-taking behavior.

Methods

This study was conducted by mail in spring 1987. Fire managers in five western regions of the USDA Forest Service participated the Northern Region, the Southwest Region, the Intermountain Region, the Pacific Southwest Region, and the Pacific Northwest Region. The regional offices provided names and addresses of all personnel in decision-making positions regarding Escaped Fire (EWF), Prescribed (prescription) Fire (RxB), and Long-Range Fire Budget Planning (LRP). Because all fire managers identified were contacted, the study is a population survey rather than a sample. The study followed the "Total Design Method" of survey research (Dillman 1978). Eighty-four percent (837 of the 994 USFS fire personnel contacted) replied with completed questionnaires.

Because environmental judgments and decisions are a product of the complex interactions among the characteristics of issues, the problem solvers, and the social and organizational context in which issues arise (Vining 1978), risk is not a generalizable trait. Rather, risk has been shown to be specific to context (Shoemaker 1980, Milburn and Billings 1976, Heimer 1988). Consistent with this literature, and to further test the context-specific nature of risk behavior, the survey instrument consisted of three modules with questions pertinent to each fire decision-making context—EWF, RxB, and LRP. To reduce respondent burden, only individuals who had experience with the particular fire-decision context received the module in that area; consequently, the number of respondents for each module was: EWF (583), RxB (633), and LRP (325). A final module asked questions about the individual respondent. Figure 1 is a schematic chart of the complete experimental design. (For discussion of the utility of using scenarios in risk research and validation of the context-specific nature of risk see Taylor et al. 1989.)

Each module began with a short, 4 - to 5-paragraph base scenario: a description of a fire situation that was as realistic as possible and which involved tradeoffs between cost and risk. For Escaped Wildfire, respondents were asked to decide whether or not to order more equipment to fight an escaped fire; for Prescribed Burning, whether to

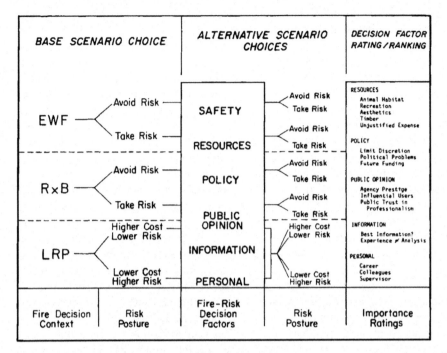

BASE SCENARIO CHOICE		ALTERNATIVE SCENARIO CHOICES	DECISION FACTOR RATING/RANKING
Fire Decision Context	Risk Posture	Fire-Risk Decision Factors	Risk Posture	Importance Ratings

Figure 1. Schematic diagram of fire-risk survey study design.

proceed with a planned burn or to postpone. Having made their base scenario choices, respondents were then directed to one of two sets of alternative scenario questions. If the response to the initial question was risk averse, the next six alternative scenario questions were designed to see what, if anything, would make the respondent move toward the more risk-taking posture. If the response to the initial scenario was risk-taking, then the next six alternative scenario questions were designed to see what, if anything, would make the respondent move toward a posture of risk-avoidance.

The risk literature suggests that risk has many dimensions, i.e., hazardous situations pose risks with different aspects or constituent parts (Hiessl and Waterstone 1986, Cortner and Waterfall 1983). Thus, in addition to being specific to context, the survey was designed to tap the different kinds of risks presented in fire management situations. Since it was hypothesized that risk

choices might be influenced bu the kinds of risk present in any one situation, information in each of the alternative scenarios reflected separately one of the following size categories of risk factors: (1) Safety; (2) Resources at risk; (3) Policy); (4) Public Opinion; (5) Information reliability; and (6) Personal considerations. A sample of the base scenario for Prescribed Burning, and two of the alternative scenario questions that use information reflecting Policy and Public Opinion risk factors are shown in Figure 2.

In all three modules, some tradeoff is being made between risk and costs. The quantitative assumption underlying all these choices was that decision-makers would have to incur higher costs to achieve lower risk levels (and vice versa). This tradeoff was made explicit in the LRP scenarios. Here, the respondent was asked to rank a series of budget/risk combinations.

Each of the alternative scenario questions altered the base scenario by changing

PRESCRIBED BURNING

PRESCRIBED BURNING-BASE SCENARIO

You are the fuels management specialist on the Summit National Forest. Your decision today is to choose whether to conduct a planned burn or to send the crew back to the station and hope for better conditions another day.

The treatment area in question is a 40-acre unit of activity fuels from a clearcutting operation. The area is scheduled to be broadcast burned by hand using a 20 member crew. The objective is to prepare the site for natural seeding of Douglas fir and to reduce the loading of slash fuel. The fuel loading is 40 tons per acre of slash which has been cured for one year. The topography is gentle, a 10% slope with a south aspect. There is a gravel road along the north side of the unit. The unit is almost square in shape with no fingerlike extensions. The adjacent stands to the south and east are Forest Service land with seedling and sapling and poletimber Douglas fir. There are no residences or other structures near. No highways are in the downwind direction whose traffic might be impeded by smoke. The weather forecast is for good burning conditions. Fuel moistures are adequate to achieve the objective and to minimize the chances of escape. You are half way through the prescribed fire season and have completed 40 percent of your planned burning.

At 1000 hours on the day of the burn, after your crew is in place and ready to start, you find that the wind is stronger and vanes in direction more than predicted. A test fire shows that burning conditions have become less desirable. The burn would still be within the allowable prescription window, but outside the preferred window since conditions are near the edge of acceptability.

You have two choices. You can postpone the burn and wait for another day when there would be less chance of escape, but you would sacrifice the cost of having set up for the fire today. The other choice is to conduct the burn today and chance an escaped fire with additional costs and resource losses. Your estimates of the consequences of each choice are shown below.

Which would you choose? (CIRCLE ONE CHOICE FROM BELOW AND GO TO PAGE INDICATED)

CHOICE A. Postpone the burn...total cost of $20,000 which includes the cost of today's set up and the cost of successfully completing the burn another day. **PLEASE GO TO PAGE PB-2**.

CHOICE B. Conduct the burn today...An 80 percent chance of no escapes and a cost of $12,000. A 20% chance of an escape either during ignition or mop up with an expected additional cost of $15,000. This total of $27,000 includes the cost of conducting the burn, controlling the escaped fire, and damages to the adjacent stand. **PLEASE SKIP TO PAGE PB-4**.

ALTERNATIVE SCENARIO A	ALTERNATIVE SCENARIO B
Change the base scenario by supposing you are ahead of schedule, having completed 60% of your burn program but only half way through the prescribed burning season. **WHAT CHOICE WOULD YOU MAKE?**	Reconsider the base scenario and suppose there is a shift in wind direction that would take smoke from the burn into a nearby community. Thus far, this community has not shown great concern about forest fire smoke. **WHAT CHOICE WOULD YOU MAKE IN THIS INSTANCE?**
CHOICE A. POSTPONE THE BURN (certain costs of $20,000)	
	CHOICE A. POSTPONE THE BURN (certain costs of $20,000)
CHOICE B. CONDUCT THE BURN (costs of $12,000 or $27,000, depending upon escape)	
	CHOICE B. CONDUCT THE BURN (costs of $12,000 or $27,000, depending upon escape)

Figure 2. Examples of RxB scenarios: base and two alternatives.

an assumption or adding a new piece of information for the respondent to consider. In both the EWF and RxB contexts, an either/or, go/no-go decision is inherent: attack with full force or with reduced force: conduct the burn or post-pone the burn. Fire budgeting, however, is done in a different kind of fire-decision context, one where there are ranges of possible choices, and more time to consider options.

Consequently, the LRP choices presented a range of annual fire budgets from $600,000 to $800,000, inversely correlated with a set of risk probabilities from 50% to 10%—of the "chance of burning more than the annual average of 2000 acres." The responses, then, are not a simple case of proceeding or not proceeding. but a ranking of these dollar/risk tradeoff choices. To yield data that could be used to compare the LRP results with the EWF and RxB results, the top three LRP choices for each respondent were aggregated into three groupings: high risk/low cost, medium risk and cost, and low risk/ high cost.

In addition to the scenario choices, an expanded list of 18-19 factors considered potentially important in making fire management decisions were given to respondents. For each of the three separate fire-decision contexts, respondents who received that module rated how important the factors were to them personally on a scale from 0 (not at all important) to 10 (extremely important).

Factors of human safety, and of protection of residences and structures were intentionally omitted from the lists of potentially important decision factors, although they were incorporated into the construction of the scenarios. It was felt that safety and protection of structures were so very important that to ask respondents to explicitly rate them might result in the other factors being rated low in comparison. That is, if fire managers anchored on a safety factor, they might rate most of the other factors at 1 or 2 in order not to detract from the stated importance of human safety.

Fire managers also *ranked* their five most influential decision factors for each decision context. Each list was keyed specifically to the fire-decision context being tested: Escaped Wildfire, Prescribed Burning, Long-Range Fire Budget Planning. Responses to these decision factor lists were analyzed and

then compared with the results of the alternative scenario decisions to give a comparison between stated importance ratings and fire-decision behavior.

To review, respondents received modules for Escaped Wildfire (EWF), Prescribed Burning (RxB), and/or Long-Range Fire Budget Planning (LRP), corresponding to their personal experience. They were asked to make an initial, base scenario decision within each module, and then a series of six alternative scenario decisions where the situation changed according to the decision factors: Safety, Resources at risk, Policy, Public Opinion, Information reliability, and Personal consideration. Finally, respondents rated the importance of some 18-19 potential fire-risk factors in influencing their decisions and ranked the top five. Separate lists of risk decision factors were evaluated by module respondents for each of the three different fire-context modules.

Results

Results from the fire manager survey are divided into three parts. First, respondents' decisions for the EWF, RxB, and LRP base scenario choices are discussed. Second, the fire managers' risk behavior, given alternative scenario choices, is analyzed to determine the influences of the different categories of decision factors on risk-taking and risk-avoiding tendencies. Finally, fire managers' ratings and ranking of items on the expanded lists of potential risk decision factors are assessed.

Base Scenario Decisions

Base scenarios were written to fit somewhere along fire managers' normal continua of risk-taking/risk-avoiding behavior. The scenarios were written, pretested and rewritten until each was judged, by fire management consultants, to be a fairly "close call" in terms of an actual situation with real choices. There is, however no a *priori* or fixed-scale means of determining just how "risk-averse"a fire manager might be in any of these contexts. Prior to this study no baseline levels for fire-risk behavior had been established. It was hypothesized, however, that given the long Forest Service history of fire suppression policy, fire managers would tend to be relatively risk-averse.

In two fire-decision contexts, EWF and RxB, respondents were given an "either or" choice. In the EWF base scenario, 63% selected Choice A, the risk averse selection to send in both crews and tankers, and 37% chose B, the more risk-taking decision to send in only crews (Table 1). For prescribed burning, Choice A, "Postpone the burn..." is the more risk-averse selection. Twenty-seven percent of those who responded to the RxB module chose this option, while 73% chose option B—the more risk-taking selection to conduct the prescribed burn (Table 1). In the LRP base scenario, five choices were given, covering a range of expenditures and an inverse range of risk probabilities. For this question, respondents were fairly evenly divided between the low, medium,and high risk choices: 35% chose the low risk/high dollar position; 29% chose

the high risk/low dollar position; and 36% selected among the mid-range choices.

Regional Differences

From the base scenario decision data it is possible to get some idea of the relative risk-taking or risk-avoiding differences among the five regions surveyed. Although the overall response cannot be put into any absolute risk-posture scale, if one group of fire managers differs importantly from the rest of the population it can be stated that that group's behavior is relatively risk-averse or risk-taking. In the Escaped Wildfire base scenario, where 37% made Choice B, the more risk-taking selection, there were some important differences among regions. Fire managers in the Southwest and Intermountain Regions were more willing to take risks in this situation, 59% and 49% respectively, while only 15% of Pacific Southwest Region fire managers were willing to accept this level of risk.

The RxB scenario differences among regions were not as great, although the Pacific Southwest people still seemed less willing to take chances with fire (37% chose to postpone compared to 27% of the total group). Fire managers in the Northern Region were

Table 1. Overall percentage of respondents who shifted risk posture in response to alternative scenarios.

Scenario	Escaped wildfire	Prescribed burning	Long-range planning
Base:	Avoid risk	Avoid risk	Low risk/high cost
	(Choice A) 63%	(Choice A) 27%	35%
	Take risk	Take risk	Medium 36%
	(Choice B) 37%	(Choice B) 73%	High risk/low cost
			29%
Shift			
Safety	74%	69%	33%
Resources	73%	45%	44%
Public	25%	38%	34%
Information	20%	37%	25%
Policy	29%	16%	7%
Personal	14%	23%	10%

slightly more prone to adopt the risk-taking posture; only 19% chose to postpone.

For LRP, fire managers in the Southwest and Intermountain Regions tended to be higher risk-takers/lower spenders, 44% and 37% respectively, compared to 29%, overall. Conversely, in the Pacific Southwest and Pacific Northwest Regions, fire managers appeared to be much more risk-averse, with 55% and 46%, respectively, selecting in the low risk/high expense choice range. Only 18% and 20%, respectively, in these latter two regions preferred the high risk choices. Fire managers in the Northern Region tended to choose the middle-of-the-road stance; 47% made choices in this range.

Overall, across the three base scenarios, Pacific Southwest Region fire managers avoided risk-taking in fire decisions more than did those from the Northern, Southwest, Intermountain, or Pacific Northwest Regions. Fire managers in the Southwest and Intermountain Regions showed greater willingness to take risks in EWF and LRP situations than the other respondents. Northern Region respondents were more willing to take risks in a Prescribed Bum situation, but not in EWF

or LRP; in the latter case these Northern Region personnel were substantially more middle-of-the-road than the others. Pacific Northwest fire managers were somewhat less willing to take risks, or more willing to spend money, in Long-Range Fire Budget Planning (but not EWF or RxB) than the others, except for the Pacific Southwest personnel.

Alternative Scenario Decisions

The first analysis of results from the alternative scenarios is the degree to which each of the six categories of decision factors-Safety, Resources at Risk, Public Opinion, Information Reliability, Policy, Personal Considerations-affected initial fire decisions. The results demonstrate that fire managers changed their risk posture as their perceptions changed of the nature of the risks involved. Table 1 shows the overall percentage of respondents who changed in response to the new information they were presented in the alternative scenarios. Tables 2 to 4 break down the direction of that change for each decision-making context and decision factor. (Direction of change is discussed later.)

Table 2. Scenario selection: escaped wildfire (EWF).

Choice A (avoid risk) crews & tankers		Scenario	Choice B (take risk) crews only	
365 [63%]		Base[a]	218 [37%]	
Avoid risk (stay with choice A)	Take risk (change to choice B)		Take risk (stay with choice B)	Avoid risk (change to choice A)
98 [27%][b]	254 [75%]	*Safety*	39 [18%][c]	180 [83%]
72 [20%]	278 [76%]	*Resources*	73 [33%]	147 [67%]
315 [86%]	38 [10%]	*Public*	111 [50%]	110 [50%]
291 [80%]	60 [16%]	*Information*	167 [77%]	54 [25%]
278 [76%]	87 [24%]	*Policy*	137 [63%]	81 [37%]
308 [84%]	44 [12%]	*Personal*	182 [83%]	37 [17%]

[a] Percent of those making original choice "A".
[b] Percent of those making original choice "B".
[c] Percentages do not total 100% because not every respondent answered for every scenario.

Changes in Risk Posture

Safety considerations and Resources at Risk had the greatest influence on fire managers' decisions. Seventy-four percent of the overall EWF, and 69% of the RxB respondents changed from their initial fire risk positions when a Safety factor, some increased danger to fire crews, was introduced. One-third of the LRP respondents shifted their initial fire-risk position when a fire-crew safety factor was added. Nearly three-quarters of the fire managers answering the EWF module, and 45% of those answering the RxB questions, changed their positions when a change was introduced vis-a-vis Resources at Risk. Forty-four percent changed from their LRP base decision grouping in response to a change in Re sources at Risk. EWF attack decisions were most susceptible to influence by Safety and Resource factors.

Third in overall sensitivity were the fire managers' responses to changes in Public Opinion, or the potential impacts a specific fire decision might have on public opinion. From one-fourth to nearly 40% of the fire managers changed from their initial decisions, influenced by changing Public Opinion factors. The LRP scenario, which changed more in response to this factor than for any other except Resources at Risk, may have done so because this alternative suggested that public concern had "the potential of escalating into a national issue."

Table 3. Scenario selection: prescribed burning (RxB).

Choice A (avoid risk) postpone		Scenario	Choice B (take risk) conduct burn	
170 [27%]		Base	463 [73%]	
Avoid risk (stay with choice A)	Take risk (change to choice B)		Take risk (stay with choice B)	Avoid risk (change to choice A)
126 [74%][a]	38 [22%][c]	*Safety*	57 [12%][b]	400 [86%]
54 [32%]	109 [64%]	*Resources*	285 [62%]	176 [38%]
119 [70%]	44 [26%]	*Public*	265 [57%]	198 [43%]
139 [82%]	22 [13%]	*Information*	245 [53%]	215 [46%]
116 [68%]	54 [32%]	*Policy*	415 [90%]	48 [10%]
144 [85%]	21 [12%]	*Personal*	334 [72%]	127 [27%]

[a] Percent of those making original choice "A".
[b] Percent of those making original choice "B".
[c] Percentages do not total 100% because not every respondent answered for every scenario.

Table 4. Scenario selection long-range planning (LRP).

Scenario	Low risk/high cost	Medium	High risk/low cost
Base	115 [(35%]	117 [36%]	93 [29%]
Safety	220 [68%]	67 [21%]	38 [12%]
Resources	255 [78%]	47 [14%]	23 [7%]
Public	225 [69%]	67 [21%]	33 [10%]
Information	194 [60%]	76 [23%]	49 [15%]
Policy	136 [42%]	101 [31%]	83 [26%]
Personal	120 [37%]	111 [34%]	33 [10%]

In RxB, which as quite sensitive to Public Opinion (38% change, overall), the possibility of also wa wind shift bringing smoke from the bum into a nearby community was suggested for the Choice B alternative scenario.

Information discrepancy (having a computer model, or in the case of RxB a weather station forecast, contradict the fire manager's experience or observations) was fourth overall in influence on fire managers' decisions. Between 20 and 40% of the respondents, varying from one fire-decision context to the next, changed their minds based on these manipulations.

Local or regional Policy changes had less influence on fire managers' decisions, ranging from a seven% to a 29% shift in fire decisions. The strongest effects, in EWF, introduced Policy statements that were in direct opposition to the initial decision: "Regional office policy has said that you should (or 'should not') use maximum aggressiveness..." The LRP Policy scenario, adding a Regional Office announcement of restricted presuppression money, seems to have elicited a response by some personnel of asking for more money when budgets are lowered, possibly as a means of establishing a tradeoff position closer to the desired budget.

Of all the decision factors introduced in these scenarios, Personal considerations had the least influence. LRP was changed very little, and only 14% of the EWF decisions were affected by considerations of impending promotion. In the RxB scenario, where the personal nature of the influence was a bit less direct the position of the new Forest Supervisor on prescribed fire-23% changed their positions vis-a-vis proceeding with the prescribed burn.

Direction of Risk Posture Change

It is of important practical and theoretical concern whether changes in risk decision posture are predominantly in one direction (e.g., toward avoiding risk) or the other. In both the EWF and RxB scenarios, two sets of alternative scenarios were offered: one set tending to push more toward taking a chance, for those who selected the risk-avoiding choice in the base scenario; and one set pushing toward avoiding risk, for those who made the more risk-taking selection initially. For the LRP scenario, the initial decision was across a spectrum of five choices, therefore only one set of alternative scenarios was provided. From three of the five choice positions within the base LRP scenario (all except the extremes), respondents could opt to move toward greater risk avoidance or greater risk taking. However, the alternative scenarios, introducing new information into the decision, by definition had to be "directional." Four of the alternative scenarios-Safety, Resources at Risk, Public Opinion, and Information Reliability-leaned toward avoiding risk; two-Policy and Personal considerations-leaned toward taking greater risk.

Safety, one of the two factors showing the greatest effect on fire managers' decision-making, showed a very strong directional influence toward risk avoidance. In the LRP Safety scenario, risk avoidance choices, gained a third (from 35 to 68%) while risk-taking lost 17% (from 29 to 12%) (see Table 4). Here, the Safety scenario stated that "fireline accidents may increase with lower budgets." In the RxB Safety scenarios (Table 3) the shift toward risk avoidance was more than 10 times stronger, in absolute numbers, than the countervailing shift. In this instance, suggesting an inexperienced crew shifted decisions toward avoidance by 86%, while having a very experienced crew only shifted decision toward risk taking by 22%. In the EWF Safety scenarios (Table 2), 83% of the B choice group shifted toward risk avoidance, sending in tankers as well as fire crews, but 75% of the

A choice group shifted toward sending in crews only. However, the latter is probably not a shift toward risk-taking so much as the avoidance of another risk, "air tankers may have trouble getting into position safely."

For Resources at Risk, second of the top two influences on fire-managers' decisions, strong shifts occurred in both directions. For both EWF and RxB the risk takers (choice B) were told that an adjacent stand contained valuable post and pole timber; 67% in EWF and 38% in RxB shifted to avoiding risk to that resource (Tables 2 and 3). However, having adjacent mature areas that had been prescribed-burned recently shifted 64% of the RxB Choice A group toward conducting the prescribed burn. When beneficial impacts to an adjacent, light-fuel stand were suggested, three quarters of the risk-avoiding fire managers in EWF shifted toward using the less intense attack. In the LRP Resource values scenario, summer residences and recreation facilities were introduced into the forest under consideration, causing an increase in risk avoidance of 43% and a 22% drop in risk-taking behavior.

Fire managers' responses to changes in Public Opinion were generally to avoid risk. After the press had covered a recent escaped fire, half changed from their base, risk-taking EWF position. (It is interesting to compare this result from the 1987 survey with the (temporary) risk aversion in prescribed natural fire in 1989, following the dramatically publicized 1988 fire season.) If smoke threatened to hit a local community, 43% of the original RxB risk takers opted out of conducting the burn. Conversely, only 10 to 26% changed to the risk-taking position in these two fire-decision contexts when public pressure was shifted in the opposite direction. In the LRP Public Opinion scenario, 34% changed to a risk-avoiding stance when public concern was expressed over increasing fire severity.

A computer model of fire behavior that didn't agree with the fire manager's experience and judgment, caused only 16 to 25% of the fire managers to shift their decisions in the EWF Information scenarios. In the LRP scenario, such a model suggesting greater risk brought 25% of the fire managers to shift to risk avoidance. When increasing wind speed and variability were predicted by a fairly remote weather station, 10 miles away, 46% of the fire managers in the RxB scenario shifted away from conducting a prescribed burn. Only one-tenth as many fire managers shifted in the opposite direction when the station predicted calmer, more stable wind conditions.

When given a regional office Policy statement in direct contradiction to their initial EWF attack choice, 24 to 37% of the fire managers altered their decisions. Interestingly, 71% of the total EWF population did not change their minds to conform to regional policy. Concerns about meeting or overshooting their prescribed burn schedules caused only 100 of the 630 RxB fire managers to change their minds, and as stated before, a tightened budget caused some Long-Range Fire Budget Planning respondents to ask for more money rather than less.

Finally, Personal decision factors had almost negligible impact on fire-decision behavior with one exception: 27% of the fire managers who decided initially to conduct the prescribed burn were swayed by their new Forest Supervisor who was "only lukewarm about prescribed burning."

In sum, of the 12 alternative scenarios in EWF and RxB that attempted to sway the fire manager toward risk avoidance, 8 succeeded in persuading over 30% of the risk-taking respondents to change their minds. Of the 12 opposite scenarios, only 4 succeeded in getting 30% or more of the risk-averse respondents to change to risk-taking. Of these, one was actually a counter-risk, and two suggested possible benefits of fire to the resource base. LRP decisions were predominantly swayed toward risk aversion, but then four of the six scenarios were weighted in that direction. The two weighted toward risk-taking, the Policy and Personal scenarios, turned out to be the two least influential decision factors overall.

Importance Ratings for Fire Risk Factors

This section discusses fire managers' responses to the decision factor importance rating exercise. Table 5 lists the seven top-rated decision factors. Reducing the chance of a catastrophic fire received the highest ranking of all the decision factors listed (7.9 on the 0-10 scale), but this factor was offered only for the LRP decision context. The remaining top-rated factors were: (1) valuable timber could be destroyed; (2) More money could be spent than is justified by the resource; (3) Public trust in agency professionalism could be damaged; (4) Future judgmental discretion could be limited; (5) Personal experience might differ from analysis or interpretation; (6) Key animal habitat could be destroyed/improved; and (7) The fire analysis given might not be based on the best available information. These ratings are quite close, with the average ratings ranging only from 6.7 to 7.9 on a 1 to 10 scale.

Aside from safety issues, which were intentionally left out of these lists, fire managers were generally most concerned about resource protection, especially timber and animal habitat. They also were highly concerned about spending too much and about maintaining public trust in the agency's professionalism-both of which could affect future decision discretion. Another top priority was getting the best possible information and analyses upon which to base their fire-management decisions.

The factors from the lists of decision factors that rated lowest in overall importance in fire management decisions are shown in Table 6. The lowest of these decision factors were: (19) One's career potential could be damaged; (18) The decision may result in reprisals from a supervisor; (17) One's decision would be under close

Table 5. The seven highest rated decision factors including [mean rating].

Overall	Escaped wildfire	Prescribed burning	Long-range planning
Timber [7.7]	Timber [7.8]	Timber [7.8]	Lower fire [7.9] chances (*)
Unjustified expense [7.5]	Unjustified expense [7.4]	Public trust [7.5]	Unjustified expense [7.6]
Public trust [7.4]	Animal habitat [7.3]	Unjustified expense [7.3]	Public trust [7.4]
Future [7.0] discretion	Public trust [7.1]	Future [7.2] discretion	Timber [7.4]
Experience vs. analysis [7.0]	Recreation [7.0]	Animal habitat [7.1]	Future [7.2] discretion
Animal [7.0] habitat	Experience vs. analysis [6.9]	Best [7.1] information	Experience vs. analysis [7.2]
Best [6.8] information	Future [6.7] discretion	Experience vs. analysis [6.9]	Schedule [7.0] (**)

(*) Category give in *only* LRP context.
(**) Category given in 2 out of 3 contexts.

scrutiny by colleagues (or subordinates); (16) Influential groups may support/be critical of the action; (15) Public anger might be aroused; (14) Aesthetic (or visual) quality might be affected; and (13) Smoke could drift into nearby communities. Again, these average ratings are fairly close.

Fire managers clearly do not express that they base their decisions on how they might look to their supervisors, their colleagues, or their subordinates. They don't feel their professional fire decisions are greatly swayed by public pressure or the reactions of influential groups. Finally, aesthetic qualities do not weigh heavily in fire-management decisionmaking.

In addition to rating each of the fire-decision factors listed, the fire managers also placed their top five factors in rank order. This helped, in the analysis of the survey data, to discriminate among the few top ratings. For example, if timber, public trust, and animal habitat all were rated "10," the rankings determined which one the manager actually considered to be the most important. Further, comparing aggregate ratings and rankings of these decision factors provides a validity check on the data.

A comparison of the top ratings and rankings for all three fire-decision contexts (Table 7), shows strong consistency in fire managers' responses on several important fire-decision factors. Protection of valuable timber resources, avoiding unjustified expenditures, retaining public trust in agency professionalism, and protecting future judgmental discretion were consistently rated and ranked as very important in all three fire-decision contexts.

Aside from these similarities, we can also look for discrepancies in importance from one context to another. Protection of animal habitat figured quite highly in both EWF and RxB ratings, third and fifth respectively (Table 7), but not at all heavily in Long-Range Fire Budget Planning where this factor was rated twelfth. The concern whether one has the best information available for decision-making figured somewhat more importantly where decisions are more immediate and cannot wait for improved information. This factor was rated sixth in RxB and eighth in EWF, but eleventh in LRP. The concern for preserving outdoor recreation opportunities was of low importance in RxB (fifteenth) and LRP

Table 6. The seven lowest rated decision factors (*) including [mean rating].

Overall	Escaped wildfire	Prescribed burning	Long-range planning
Career [4.7]	Supervisor [4.3]	Career [5.0]	Colleagues [4.6]
Supervisor [4.8]	Influential groups [4.3]	Supervisor [5.1]	Career [4.8]
Colleagues/subord. [4.9]	Career [4.4]	Subordinates [5.6]	Performance rate (Super) [5.1]
Influential grups [5.3]	Colleagues [4.6]	Recreation [5.8]	Influential groups [5.3]
Public (**) anger [5.6]	Smoke [4.7] in community	Aesthetics [5.9]	Aesthetics [5.5]
Aesthetics [5.7]	Budgets [5.0]	Public (**) anger [6.1]	Recreation [6.2]
Smoke [5.7] in community	Public (**) anger [5.1]	Schedule (**) [6.1]	Animal [6.4] habitat

(*) Lowest rated listed first.
(**) Category given in only 2 out of the 3 contexts.

(thirteenth) decisions, but quite important in EWF decisions (fifth). The concern that smoke might get into nearby communities was much more important when managers were setting the fire (ninth rated factor in RxB) than for fires not set by managers (EWF where it was rated fourteenth).

Discussion

The results of this study illustrate that risk behavior and perceptions vary by region and by fire-decision context. Any particular risk decision-making situation is characterized by multiple dimensions and by individual assessments of the relative importance of risk factors. The complex influences on fire-risk behavior suggest that "risk" has often been too narrowly conceptualized. Risk has been traditionally treated as gambler's odds or probability. In decision-making contexts, however, risk includes probabilities, values at risk, and costs of avoiding. This research affirms that "perception of risk" is influenced by situational factors, and that risk in a decision-making context is a multidimensional perceptual product. Risk attitudes and perceptions affect how managers respond to current as well as to emerging agency fire management policies.

Regional Variation

Some of the identified regional differences may well have to do with variations in the fire regimes of different forest types. It could be quite logically predicted that fire managers from California, where some national forests contain fire-prone chaparral ecosystems, would be much more cautious in their approach to fires of all sorts. To some degree, the fire management climate must reflect fire regime. However, the distinct regional differences identified here are important enough to warrant agency discussion of whether those differences are desirable or within the acceptable range of variability for a decentralized organization.

Risk and the Urban/Wildland Interface

Another explanation of these differences may be related to differences, from region to region, in the degree of interspersion of urban development within or adjacent to wildlands under the Forest Service's jurisdiction. The problem to date has been largely perceived as a southern California problem-created by the interaction of extensive chaparral ecosystems and 14 million people. But increasingly, the issues brought about by the interaction of residential

Table 7. Comparisons of factor ratings and rankings.

Factor:	Escaped wildfire		Prescribed burning		Long-range planning	
	Rating	Ranking	Rating	Ranking	Rating	Ranking
Lower fire chances (*)					1	1
Timber	1	2	1	1	4	6
Unjustified expense	2	1	3	2	2	2
Public trust	4	5	2	4	3	3
Future discretion	7	4	4	3	5	4
Experience vs. analysis	6	—	7	—	6	—
Animal habitat	3	3	5	—	12	—
Best information	8	—	6	5	11	—
Meet schedule objective (**)			13	—	7	5
Political problems	9	—	8	—	8	—
Smoke in community (**)	14	—	9	—		

(*) Category given in LRP context only.
(**) Category given in only 2 of the 3 contexts.

development in or near forests are occurring throughout the West, indeed throughout the United States. Problems are arising in three distinct areas (1) at the fringe of densely populated metropolitan areas; (2) in the ex-urbs beyond large metropolitan areas; and (3) in remote rural areas, largely in the West (Shands 1988, p. 6–7).

That the urban-wildland fire interface is important in fire-risk behavior is strongly supported by the Long-Range Fire Budget Planning results, where the greatest shift toward risk avoidance was in response to residences and recreational facilities being introduced into the forest in question.

The results have important behavioral implications for the Forest Service's newly emerging concern about problems related to the interface. As the inter mixing of structures and wildlands increase, there is likely to be greater public pressure to commit more and more resources to preventing, fighting, and mitigating wildland fires that threaten or appear to threaten developed areas. The size, training, and equipping of the agency's firefighting forces may be influenced by concerns about the risks posed by the urban-wildland interface fire hazard. Since most long-range fire budget choices involve explicit trade-offs between efficiency and risk, the increased pace of interface development could also en gender closer attention to structural values in determining fire efficient budgeting. Politically, management responses will not only need to be responsive to the biophysical and socioeconomic dimensions of the fire problem, but also to both public and agency perceptions of the nature and magnitude of the risks involved.

Multiple Dimensions Of Risk

The two primary categories of factors affecting fire managers' decisions to take or avoid risk are Safety and Resources at risk

Fire managers can be substantially swayed from their original course of action when they discover a potential threat to human safety or a new safety risk countervailing their original safety concerns. This reaffirms that there is strong support for the long-standing Forest Service policy that the first item of importance in fire fighting is safety.

Secondly, when Resources at Risk are threatened, fire managers' decisions can be substantially influenced, but the specific type of resource change is critical. Added threats to aesthetic resources do not weigh heavily on fire-risk decisions, nor do threats to recreation resources for prescribed fire or fire budget planning. On the other hand, putting timber stands at risk or spending more for suppression than the resources warrant rated high as an influence on fire managers' decisions. Introduction of summer residences and facilities greatly swayed decisions on Long-Range Fire Budget Planning, more than any other factor. This latter scenario, however, may have been felt to involve public safety as well as threats to resources. Thus, fairly consistently, those resource values which swayed decisions were directly tied to market products; non-market economic values, such as recreation and aesthetics, held considerably less sway. This finding is consistent with recent studies which show a high Forest Service commitment to traditional commodity production and less commitment to other multiple-use activities and environmental values (Twight and Lyden 1988).

Public trust, which was rated high in all three fire-decision contexts, is obviously a Public Opinion-related factor, but so are two of the lowest rated decision factors: influential groups might be critical, and public anger might be aroused. The public trust factor included the very important phrase "in agency professionalism" which was not part of lower rated Public Opinion factors.

Fire managers also seemed reluctant to say they would be influenced by public anger or by influential groups, which further supports their perception of their agency as a highly professional organization. On the other hand, managers were sensitive to possible community concerns with smoke from a prescribed burn and to the possibility of an issue escalating into a national issue. These findings corroborate the conventional wisdom that there is considerable agency sensitivity to community concerns about smoke impacts from prescribed bums. The findings also suggest that unqualified use of the term "public opinion" in the forestry and risk behavior literature may be too broad to be meaningful. It would appear that this term encompasses some decision factors considered to be very important and others which are intentionally ignored or down-played.

Two decision factors which are conceptually related to the Information reliability scenarios are "experience vs. analysis" and "best information." Although the Information scenarios effected little change in fire managers' decisions, these factors rated high. In each of the information scenarios, some other information system-computer model, weather station-disagreed with the experience and judgment of the fire manager or his/her staff. Fire managers appear to trust their own knowledge and experience, which may be related to "future discretion" being rated so high as a decision factor.

Turning to those factors displaying the least influence on fire managers' risk decisions, Personal considerations emerge as the least influential on decisions both in factor ratings and in decision scenarios. Having a decision under scrutiny by a supervisor or colleague, or the potential for a decision to influence an impending promotion, had virtually no effect on fire managers' risk decisions in the scenarios. Further, ratings for decision factors such as career promotion, super visor disapproval, or critical review

by colleagues or subordinates were among the lowest. There can be no question that professional foresters, as people in any profession, must bear in mind their chances of promotion, as well as their super visors' and colleagues' opinions of their work. We shall never know whether the responses we received on these decision factors truly reflect these professionals' innermost feelings or whether, through training and socialization, these are the only acceptable responses. What is noteworthy is that fire managers are quite consistent in stating that these factors do not influence individual fire-risk decisions, neither through scenario decisions nor in factor ratings.

Having a regional policy stated that is counter to one's original decision, or having a restrictive budgetary policy, had little impact on fire managers' decisions. Once a fire management risk decision is made as to the wisest course of action, immediate or local changes in policy are not likely to sway that course. This is not to say that agency policy does not weigh importantly among fire managers. Con sider the importance given to the factor, "more money could be spent than is justified by the resources saved," which ranked second, overall, of all decision factors in influencing fire decisions. This balance between resource values and fire costs represents a fundamental policy change instituted over the past decade within the Forest Service. Clearly, fire managers see themselves taking this relatively new policy directive quite seriously.

Changing Managers' Behavior

The results of this study also have implications for the kinds of incentives and rewards used to motivate Forest Service personnel to adopt new fire management policies and practices. The findings show that fire management professionals would be slow to change their behavior in response to new directives of fire management policy without mechanisms to relate this to on-the-ground choices, e.g., decision rules or standards. Reprisals for noncompliance would be even less likely to effect change. On the other hand, demonstrating that a new policy or direction would result in increased safety to both people and resources and would retain options for professional decisionmaking, should assure change in fire decision behavior much more quickly, and by a significant proportion of fire managers. This underscores the importance of how incentives and inducements for effecting change in behavior are framed (Tversky and Kahneman 1981, Hogarth 1980).

Conclusion

Overall, study results found that fire managers appear to be susceptible to pressures to avoid risk in fire decision-making processes. Fire-risk behavior varied from one geographic fire regime to another and from one fire-decision context to the next. and while certain categories of decision factors, primarily Safety and Resources at risk had a much stronger effect on fire-risk decisions than other factors, these influences were not uniform. Managers were differentially sensitive to a number of issues-safety, resources at risk balanced with costs, the public's view of agency professionalism as opposed to pressure group politics, agency policy vs. local and regional directives, the quality of information and information models, as well as the ability to maintain judgmental discretion for management professionals. Information on fire-risk decision behavior should be weighed care fully by anyone-within or without the Forest Service-who might wish to sway fire decisions or form fire policy in the future.

Literature Cited

Bolgiano, C. 1989. Yellowstone and the Jetburn policy. Arn. For. 95(1&2):21–25.

Cortner, H. J., and P. H. Waterfall. 1983. Measuring risk preferences in fire management situations. Proj. completion rep. USDA For. Serv., Pac. Southwest For. Range Exp. Stn

Dillman, D. A. 1978. Mail and telephone survey: The total design method. Wiley, New York.

Fire Management Policy Review Team. 1988. Report on fire management policy. Dep. of Agric.,Dep. of Interior, Washington, DC.

Gale, R. D. 1977. Evaluation of fire management activities on the national forests. USDA For. Serv. Pol. Anal. Staff, Washington DC.

Heimer, C. A. 1988. Social structure, psychology, and the estimation of risk. Annu. Rev. Soc. 14:491-519.

Hiessl, H., and M. Waterstone. 1986. Issues with risk. Univ. of Arizona Water Resour. Res. Cent., Tucson.

Hogarth, R. M. 1980. Judgment and choice: Strategies for decision. Wiley, New York.

Kilgore, B. M. 1984. Restoring fire's natural role in America's wilderness. West. Wildl. 10(3):2–8.

Lewin, R. 1988. Ecologists' opportunity in Yellowstone blaze. Science 241:1762–1763.

Milburn, T. W., and R. S. Billings. 1976. Decisionmaking perspectives from psychology. Arn. Behav. Sci. 20(1):111-126.

Nelson, T. C. 1979. Fire management policy in the national forests-a new era. J. For. 77(1):723–725.

Shands, W. E. 1988. Forest wildlands and their neighbors: interactions, issues, opportunities. USDA For. Serv., Washington, DC.

Shear, J. 1988. So Jong, Smokey, old buddy; natural way is the fire policy. Nation (August 29):18–20.

Shoemaker, P. J. H. 1980. Experiments on decisions under risk: The expected utility hypothesis. Martinus Nijhoff, Boston.

Taylor, J. G., et al. 1989. Risk perception and behavioral context: US Forest Service Fire Management Professionals. Soc. Natur. Resour. 1:253-268.

Tversky, A., and D. Kahneman. 1981. The framing of decisions and the psychology of choice. Science 211:453–458.

Twight, B., and F. Lyden. 1988. Multiple use v. organizational commitment. For. Sci. 34(2):474–486.

Vining, J. 1987. Environmental decisions: The interaction of emotions, information, and decision context. J. Environ. Psych. 7:13–30.

Wakimoto, R. H. 1989. National fire management policy: A look at the need for change. West. Wildl.15(2):35–39.

Authors and Acknowledgments

Hanna J. Cortner is Professor, Renewable Natural Resources, and Associate Director, Water Resources Research Center, University of Arizona, Tucson 75721; Jonathan G. Taylor, Environmental and Societal Impacts Group, National Center for Atmospheric Research, Boulder, CO (Current affiliation: Research Social Scientist, National Ecology Research Center, U.S. Fish and Wildlife Service, Fort Collins, CO 80526); Edwin H. Carpenter, is Professor, Department of Agricultural Economics, University of Arizona, Tucson 85721; and David A. Cleaves is Associate Professor, Forest Management Department, Oregon State University, Corvallis 97331.

Part 3

Fuels Management: Forestry Meets Fire

■ Brian Van Winkle

A ny anthology of fire management research produced by Society of American Foresters (SAF) must consider its body of available literature on the subject of fuels, given the veritable Augean Stable status of today's fuels management issues. To do otherwise would be intellectually criminal. Across the multifaceted realm of fire management the management of fuels holds a special place for the forester, for what is fuels management if not the manipulation of vegetation (both amount and arrangement), which is the special calling and professional provenance of the forester? If one looks at the basic teaching tool of fire management, the fire behavior triangle (topography, weather and fuels) we see that the fuels management side is the only one that human intervention can meaningfully impact. Cloud seeding and mountaintop removal aside, we recognize

that our real contribution to how fires will burn is through the practice of vegetation manipulation that is fuels management.

The papers selected for this topic area, chosen from the hundreds published over the last 115 years in SAF's journals, represent the most significant contributions to the subject of fuels management. However, when considering the above statement, one must ask the question: "what constitutes a significant contribution?" There are, of course, a number of ways to consider the significance of scientific literature. One is to consider a work's impact in its own time and context. Another is to trace the role of the work in inspiring and building future works. Yet another is to consider the relevance of the work today given current advances, scientific understanding and objectives. Considering both the sheer diversity of the selected papers and the time span that they encompass,

I will limit my comments to a few statements about each one individually, giving the reader the barest topographic outline while allowing the works to speak for themselves. Though these papers all had some level of impact in their time and each has been cited multiple times since its publication, for the sake of space I will only make comments based on the aforementioned third criteria (relevance under the current management regime).

The first paper, "Grass, Brush, Timber and Fire in Southern Arizona" by Aldo Leopold ties together the encroachment of woody brush, reductions in forage production and widespread erosion with post-Anglo settlement shifts in fire regimes (most often cessation of fire). Considering the millions of acres of land, both public and private, that are currently affected by this kind of encroachment across multiple ecosystems (grasslands, savannas, sage-steppe, Pinyon-Juniper woodland, and so on), this may be one of the most currently relevant issues in fuels management today. Leopold's conclusions and methods stand the test of time. A blending of social science and human dimensions techniques (such as oral histories and archival research) with some proto-dendrochronology and field observations give the work an almost palpable modernity.

The second paper, "The Fuels Buildup in American Forests" by Carl Wilson and John Dell, discusses the issue of fuel accumulation, options for dealing with these fuels, and what research is needed to effectively expand treatments. With respect to fuel accumulation the authors refine the discussion by separating this into "nature's contribution" (in situ fuels) and "man's contribution" (primarily activity fuels). The authors put the costs of a sampling of selected major fires of the era into perspective in terms of their financial costs, their acreages, and the loss of structures and lives. But for the

specific fires mentioned within, this paper reads like one written today. This is a testament to the maxim that "the more things change, the more they stay the same." Do take note of one favorite, and possibly inadvertent, comment on the folly of a one-size-fits-all (ecosystems or socioeconomic systems) fire management approach.

The third paper, "Problems and Priorities for Forest Fire Research" by Craig Chandler and Charles Roberts, makes particular comment on the need for practical and timely fire and fuels management research. The authors propose four focus areas for future fire research: 1) hazard reduction, 2) fire behavior and effects, 3) risk reduction and 4) technology development (and, by extension, its transfer). Considering that these categories still form the backbone of fire research today, this paper too stands up well in the current fuels management discussion. A cursory search of fire science publications and funded research projects over the last two decades will yield few that don't fit into one or more of these categories. Whereas the realm of scientific research is always evolving to meet mankind's newest challenges, it seems inconceivable that these categories of inquiry will change any time soon.

In the fourth paper, "Time for a New Initiative," authors Stephen Arno and James Brown propose zoning National Forests and other units of public land by differing levels of development into what is today called the Wildland Urban Interface (or WUI for short). They suggest that each corresponding zone have a prescribed set of options available or, conversely, not available, specifically with respect to the application of fire, whether it be prescribed or lightning caused. They also include mechanical surrogates such as timber harvest. This concept meshes almost perfectly with the modern concept of the "Red–Green Map" used by public land management agencies. Whereas current uses of this tool carefully avoids being predecisional, they recommend zone managed land by what suite of options are available therein. These include the full spectrum of fire and fuels management from aggressive "full suppression" to "building a large box" for fires managed with less aggressive containment strategies. All of this is done with the objective of reaping the potential benefits of having more fire on the land. The perceived benefits span the gamut from less intense wildfire as a result of a reduced fuel load to greater decision space for fire managers and even to the full suite of ecosystem services, such as species composition and nutrient cycling.

The last paper, "Using Fire to Increase the Scale, Benefits and Future Maintenance of Fuel Treatments" by Malcolm North, Brandon Collins and Scott Stephens, explores the opportunities presented by the Forest Service's 2012 planning rule to increase the acreage of National Forest System Lands treated with fire by incorporating fireshed objectives directly into Forest Land and Resource Management Plans. Seeing as this work is only five years old, it is obviously relevant to today's issues and the current level of scientific understanding. It is perhaps best left to future generations to consider how much of a precedent this work may or may not have had and whether it stands the test of time as the others included in this collection have. Its inclusion in this anthology can best be characterized as a possible look to the future by way of the present, taking what is currently considered a good idea that may impact the future and betting that it will.

As well as having impact at the time of publication and influencing further works,

each of these papers stand up well under the current management and science environment—indeed, some uncannily so. Topics like fire history, woody vegetation encroachment, the impacts of grazing, soil erosion, watershed degradation, fuel load accumulation, suppression costs, biomass utilization, treatment efficacy, research priorities, hazard reduction, fire behavior and effects, risk reduction, technology development, urban-interface conflicts, smoke management and associated liability, public outreach and education, fireshed planning: all of these facets are contained in the selections, and all are still central to the discussion of fuels management today.

If those who ignore history are doomed to repeat it, then perhaps those who study the works of the past will be in the best position to make progress toward society's goals while avoiding the bumps in the road we've previously hit. While foresters will benefit from the information contained in the papers in this anthology, its greatest contribution may be as an illustration that there is, as they say, nothing new under the sun. The fuels management issues of today are either the same as, parallel to, or heavily influenced by those of the past. This being the case, we in the field of forestry, fire and land management have a tool our predecessors lacked: the totality of their work. This may be the key to cleaning our Augean Stable, or at least meaningfully sprucing it up. To tackle this work, we as a profession will need the broadest perspective possible, not just across scientific disciplines, but across the gulf of years as well. This chapter is a timely first step in assembling works with just such a perspective.

Grass, Brush, Timber, and Fire In Southern Arizona

1924. *Journal of Forestry*. Vol 22. No. 6. 1–10.

Aldo Leopold
U.S. Forest Service

One of the first things which a forester hears when he begins to travel among the cow-camps of the southern Arizona foothills is the story of how the brush has "taken the country." At first he is inclined to classify this with the legend, prevalent among the old timers of some of the northern states, about the hard winters ·that occurred years ago. The belief in the encroachment of brush, however, is often remarkably circumstantial. A cow-man will tell about how in the 1880's on a certain mesa he could see his cattle several miles, whereas now on the same mesa he can not even find them in a day's hunt. The legend of brush encroachment must be taken seriously.

Along with it goes an almost universal story about the great number of cattle which the southern Arizona foothills carried in the old days. The old timers say that there is not one cow now where there used to be 10, 20, 30, and so on. This again might be dismissed but for the figures cited as to the brandings of old cattle outfits, of which the location and area of range are readily determinable. This story likewise must be taken seriously.

In some quarters the forester will find a naive belief that the two stories represent cause and effect, that by putting more cattle on the range the old days of prosperity for the range industry might somehow be restored.

The country in which the forester finds these prevalent beliefs consists of rough foothills corresponding in elevation to the woodland type. Above lie the forests of western yellow pine, Below lie the semi-desert ranges characteristic of the southern Arizona plains. The area we are dealing with is large, comprising the greater part of the Prescott, Tonto, Coronado, and Crook National Forests as well as much range outside the Forests. The brush that has "taken the country" comprises dozens of species, in which various oaks, manzanita, mountain mahogany and ceanothus predominate. Here and there alligator junipers of very large size occur. Along the creek bottoms the brush becomes a hard wood forest.

Five facts are so conspicuous in this foothill region as to immediately arrest the attention of a forester.

1. Widespread abnormal erosion. This is universal along watercourses with sheet erosion in certain formations, especially granite.
2. Universal fire scars on all the junipers, oaks, or other trees old enough to bear them.
3. Old juniper stumps, often leveled to the ground, evidentially by fire.
4. Much juniper reproduction merging to pine reproduction in the upper limits of the type.
5. Great thrift and size in the junipers or other woodland species which have survived fire.

A closer examination reveals the following additional facts:

First, the reproduction is remarkably even aged. A few ring counts immediately establish the significant fact that none of it is over 40 years old. It is therefore contemporaneous with settlement; this region having been settled and completely stocked with cattle in the 1880's.

Second, the reproduction is encroaching on the parks. These parks, in spite of heavy grazing, still contain some grass. It would appear, therefore, that this reproduction has something to do with grass.

Third, one frequently sees manzanita, young juniper or young pines growing within a foot or two of badly fire-scarred juniper trees. These growths being very susceptible to fire damage, they could obviously not have survived the fires which produced the scars. Ring counts show that these growths are less than 40 years old. One is forced to the conclusion that there have been no widespread fires during the last 40 years.

Fourth, a close examination of the erosion indicates that it, too, dates back about 40 years and is therefore contemporaneous with settlement, removal of grass, and cessation of fires.

These observations coordinate themselves in the following theory of what has happened: Previous to the settlement of the country, fires started by lightning and Indians kept the brush thin, kept the juniper and other woodland species decimated, and gave the grass the upper hand with respect to possession of the soil. In spite of the· periodic fires, this grass prevented erosion. Then came the settlers with their great herds of livestock. These ranges had never been grazed and they grazed them to death, thus removing the grass and automatically checking the possibility of widespread fires. The removal of the grass relieved the brush species of root competition and of fire damage and thereby caused them to spread and "take the country." The removal of grass-root competition and of fire damage brought in the reproduction. In brief, the climax type is and always has been woodland. The thick grass and thin brush of pre-settlement days represented a temporary type. The substitution of grazing for fire brought on a transition of thin grass and thick brush. This transition type is now reverting to the climax type—woodland.

There may be other theories which would coordinate these observable phenomena, but if there are such theories nobody has propounded them, and I have been unable to formulate them.

One of the most interesting checks of the foregoing theory is the behavior of species like manzanita and pinon. These species are notoriously susceptible to fire damage at all ages. Take manzanita: One finds innumerable localities where manzanita thickets are being suppressed and· obliterated by pine or juniper reproduction. The particular manzanita characteristic of the region (Arctostaphylos pungens) is propagated by brush fires, seedling (not coppice) reproduction taking the ground whenever a fire has killed the other brush species or reduced them to coppice. It is easy to think back to the days when these manzanita thickets, now being killed, were first established by a fire in what was then grass and brush. Cattle next removed the grass. Pine and juniper then reproduced due to the absence of grass and fire, and are now overtopping the manzanita. Take pinon: It is naturally a component of the climax woodland type but mature pinons are hardly to be found in the region; just a specimen here and there sufficient to perpetuate the species which has evidently been decimated through centuries of fires. Nevertheless today there is a large proportion of pinon in the woodland reproduction which is coming in under some of the Prescott brushfields.

Another interesting check is found in the present movement of type boundaries. Yellow pine is reproducing down hill into the woodland type. Juniper is reproducing down hill into the semi-desert type. This down-hill movement of type lines is so conspicuous and so universal as to establish beyond a doubt that the virgin condition previous to settlement represented a temporary type due to some kind of damage, and completely refutes the possible assumption that the virgin conditions were climax and the present tendency is away from rather than toward a climax.

A third interesting check is found in the parks. In general there are two alternative hypotheses for Southwestern parks—the one assuming chemical or physical soil conditions unfavorable to forests and the other assuming the exclusion of forests by damage. When the occasional forest tree found in any park is scrubby, it indicates in general defective soil conditions. When the occasional forest tree shows vigor and thrift, it indicates that the park was established by damage and that the soil is suitable. Nothing could be more conspicuous than the vigor

and thrift of the ancient junipers scattered through the parks of the southern Arizona foothills. We may safely assume that these parks were not caused by defective soil conditions. That they were caused by grass fires is evidenced by the survival of grass species in spite of the extra heavy grazing which occurs in them and by the universal fire scars that prevail on the old junipers in them. The fact that they are now reproducing to juniper clinches the argument.

A fourth check bears on the hypothesis that the virgin grass was heavy enough to carry severe fires. The check consists in the occurrence of "islands" where topography has prevented grazing. One will find small benches high on the face of precipitous cliffs which, in spite of poor and dry soil, bear an amazing stand of grasses simply because they have never been grazed. One even finds· huge blocks of stone at the base of cliffs where a little soil has gathered on the top of the block and a thrifty stand of grasses survives simply because livestock could not get at it.

The most impressive check of all is the occurrence of junipers evidently killed by a single fire from 50 years to many centuries ago, on areas where there is now neither brush nor grass and where the junipers were so scattered (as evidenced by their remains) that it is absolutely necessary to assume a connecting medium. If the connecting medium had been brush it could hardly have been totally wiped out because neither fire nor grazing exterminates a brushfield. It is necessary to assume that the connecting medium consisted of grass. It is significant that the above described phenomenon occurs mostly on granitic formations where it is easy to think that a heavy stand of grass might have been exterminated by even moderate grazing due to the loose nature of the soil.

Assuming that all the foregoing theory is correct, let us now consider what it teaches us about erosion. Why has erosion been enormously augmented during the last 40 years? Why has not the encroachment of brush checked the erosion which was induced by the removal of the grass? Why did not the fires of pre-settlement days cause as much erosion as the grazing of post-settlement days?

It is obvious at the start that these questions can not be answered without rejecting some of our traditional theories of erosion. The substance of these traditional theories and the extent to which they must be amended before they can be applied to the Southwest, I have discussed elsewhere.[1] It will be well to repeat, however, that the acceptance of my theory as to the ecology of these brushfields carries with it the acceptance of the fact that at least in this region grass is a much more effective conserver of watersheds than foresters were at first willing to admit, and that grazing is the prime factor in destroying watershed values. In rough topography grazing always means some degree of localized overgrazing, and localized overgrazing means earth-scars. All recent experimentation indicates that earth-scars are the big causative agent of erosion. An excellent example is cited by Bates, who shows that the logging road built to denude Area B at Wagon Wheel Gap has caused more siltage than the denudation itself. Another conspicuous example is on the GOS cattle range in the Gila Forest, where earth-scars due to concentration of cattle along the water-courses have caused an entire trout stream to be buried by detritus, in spite of the fact that conservative range management has preserved the remainder of the watershed in an excellent condition.

Let us now consider the bearing of this theory on Forest administration. We have) earned that during the pre-settlement period of no grazing and severe fires, erosion was not abnormally active. We have learned that during the post-settlement period of no fires and severe grazing, erosion became exceedingly active. Has our administrative policy applied these facts?

It has not. Until very recently we have administered the southern Arizona Forests on the assumption that while overgrazing was bad for erosion, fire was worse, and that therefore we must keep the brush hazard grazed down to the extent necessary to prevent serious fires.

In making this assumption we have accepted the traditional theory as to the place of fire and forests in erosion, and rejected the plain story written on the face of Nature. He who runs may read that it was not until fires ceased' and grazing began that abnormal erosion occurred. We have likewise rejected the story written in our own fire statistics, which shows that on the Tonto Forest only about 1/3 of 1% of the hazard area burns over each year, and that it would therefore take 300 years for fire to cover the forest once. Even if the more conservative grazing policy which now prevails should largely enhance the present brush hazard by restoring a little grass, neither the potential danger of fire damage nor the potential cost of fire control could compare with the existing watershed damage. Moreover the reduction of the brush hazard by grazing is to a large degree impossible. This brush that bas "taken the country" consists of many species, varying greatly in palatability. Heavy grazing of the palatable species would simply result in the unpalatable species closing in, and our hazard would still be there.

There is one point with respect to which both past policy and present policy are correct, and that is the paramount value of watersheds. The old policy simply erred in its

---

1. "A Plea for Recognition of Artificial Works in Forest Erosion Control Policy," *Journal of Forestry*, March, 1921. "Pioneers and Gullies," *Sunset Magazine*, May, 1924. *Watershed Handbook*, Southwestern District, issued December, 1923.

diagnoses of how to conserve the watershed. The range industry on the Tonto Forest represents a present capital value of around three millions. Since this is about one third of the total Roosevelt Reservoir drainage we may assume roughly that the range industry affecting the Reservoir is worth nine millions. The Roosevelt Dam and the irrigation works of the Salt River Valley represent a cash expense by the Government of around twelve millions. The agricultural lands dependent upon this irrigation system are worth about fifty millions, not counting dependent industries. Grazing interests worth nine millions, therefore, must be balanced against agricultural interests worth sixty-two millions. To the extent that there is a conflict between the existence of the range industry . and the permanence of reclamation, there can be no doubt that the range industry must give way.

In discussing administrative policy, I have tried to make three points clear: First, 15 years of Forest administration were based on an incorrect interpretation of ecological facts and were, therefore, in part misdirected. Second, this error of interpretation has now been recognized and administrative policy· corrected accordingly. Third, while there can be no doubt about the enormous value of European traditions to American forestry, this error illustrates that there can also be no doubt about the great danger of European traditions to American forestry; this error also illustrates that there can be no doubt about the great danger of European traditions uncritically accepted and applied, especially in such complex fields as erosion.

The present situation in the southern Arizona brushfields may be summed up administratively as follows:

1. There has been great damage to the watershed resources.

2. There has been great benefit to the timber resources.

3. There has been great damage to the range resources.

Whether the benefit to timber could have been obtained with lesser damage to watersheds and ranges is an academic question dealing with bygones and need not be discussed. Our present job is to conserve the benefit to timber and minimize the damage to watershed and range in so far as technical skill and good administration can do it. Wholesale exclusion of grazing is neither skill nor administration, and should be used only as a last resort. The problem which faces us constitutes a challenge to our technical competency as foresters—a challenge we have hardly as yet answered, much less actually attempted to meet. We are dealing right now with a fraction of a cycle involving centuries. We can not obstruct or reverse the cycle, but we can bend it; in what degree remains to be shown.

There are some interesting sidelights which enter into the foregoing discussions but which could not there be covered in detail. One of them is the extreme age of the junipers and juniper stumps. In one case I found a 36" alligator juniper with over half its basal cross-section eaten out by fire. On each edge of this huge scar were four overlapping healings. The last healing on each edge of the scar counted forty rings. Within 24" of the scar were two yellow pines of 20" diameter just emerging from the blackjack stage. Each must have been 130 years old. Neither showed any scars, but upon chopping into the side adjacent to the juniper, each was found to contain a buried fire-scald in the fortieth ring. It was perfectly evident that these 130-year pines had grown in the interval between the fires which consumed half the basal cross-section of the juniper, and the subsequent fires which resulted in the latest series of four healings.

The fires which really ate into the juniper would most certainly have killed any pine standing only 24" distant. The conclusion is that the juniper attained its present diameter more than 130 years ago. The size of the main scar certainly indicates a long series of repetitions of scarring, drying and burning at the base of the juniper. The time necessary to attain a 36" diameter is in itself a matter of centuries. Consider now that other junipers killed by lire 40 years ago were found to still retain ¼" twigs, and then try to interpret in terms of centuries the meaning ·of the innumerable stumps of juniper (the wood is almost immune to decay) which dot the surface of the Arizona foothills. Who can doubt that we have in these junipers a graphic record of forest history extending back behind and beyond the Christian era? Who can doubt that this article discloses merely the main broad outlines of the story?

The following instance also tells us something about the intervals at which fires occurred. I mentioned a juniper with a big scar and four successive healings of which the last counted forty rings. The last was considerably the thickest. In a general way I would say that the previous fires probably occurred at intervals of approximately a decade. Ten years is plenty of time for a lusty growth of grass to come back and accumulate the fuel for another fire. This would reconcile my general theory with the known fact that fires injure most species of grass, it being entirely thinkable for the grass to recover from any such injury during a ten-year interval.

The foregoing likewise strengthens the supposition that root competition with grass rather-than fire, was the salient factor in keeping down the brush during pre-settlement days. Brush species which coppice with as much vigor as those of the Arizona brushfields could stage quite a comeback during a ten-year surcease of lire

if they were not inhibited by an additional competitor like grass roots.

Whether grass competition or fire was the principle deterrent of timber reproduction is hard to answer because the two factors were always paired, never isolated. Probably either one would have inhibited extensive reproduction. In northern Arizona there are great areas where removal of grass by grazing has caused spectacular encroachment of juniper on park areas. But here again both grass competition and fire evidently cause the original park, and both were removed before reproduction came in.

It is very interesting to compare what has happened in the woodland type with what has happened in the semi-desert type immediately below it. Here also old timers testify to a radical encroachment of brush species like mesquite and cat's-claw. They insist, however, that while this semi-desert type originally contained much grass, it never contained enough grass to carry fire. There are no signs of old fires. The encroachment of brush in this type can therefore be ascribed only to the removal of grass competition.

There are many loose masonry walls of Indian origin in the headwaters of drainages both in the woodland and semi-desert types. These have been fondly called "erosion-control works" by some enthusiastic forest officers, but it is perfectly evident that they were built as agricultural terraces, and that their function in erosion control was accidental. It is significant that any number of these terraces now contain heavy brush and even timber. Since they are prehistoric, the Indians could not have had metals, and therefore could not have easily cleared them of timber or brush. Therefore their sites must have been either barren or grassy when the Indians built them. This conforms with the belief that brush has encroached in both the woodland and semi-desert ranges.

In the brush fields of California the drift of administrative policy is toward heavy grazing as a means of reducing fire hazard. If the ecology of these California brushfields is similar to the ecology of the Arizona brushfields, it would appear obvious that either my Arizona theory or the California grazing policy is wrong. The point is that there is no similarity. The rainfall of the California brushfields is nearly twice that of the Arizona brushfields. Its seasonal distribution is different, and from what I can learn there is a great deal more duff and more herbs and other inflammable material under the California brush. It would appear, therefore, that the California tendency toward heavier grazing and the tendency in the Southwestern District toward much lighter grazing are not inconsistent because the two regions are not comparable.

The radical encroachment of brush in southern Arizona has had some interesting effects on game. There is one mountain range on the Tonto where the brush has become so thick as to almost prohibit travel, and where a thrifty stock of black bears have established themselves. The old hunters assure me that there were no black bears in these mountains when the country was first settled. It is likewise a significant fact that the wild turkey has been exterminated throughout most of the Arizona brushfields, whereas it has merely been decimated further north. It seems possible that turkeys require a certain proportion of open space in order to thrive. Plenty of open spaces originally existed, but the recent encroachment of brush has abolished them, and possibly thus made the birds fall an easier prey to predatory animals.

The cumulative abnormal erosion which has occurred coincident with the encroachment of brush and the decimation of grass naturally has its worst effect in the siltage of reservoirs. The data kept by Southwestern reclamation interests on siltage of reservoirs is regrettably inadequate, but it is sufficient to indicate one salient fact, viz., that the greater part of the loosened material is at the present time in transit toward the reservoir, rather than already dumped into it. Blockading this detritus in transit is therefore just as important as desilting the storage sites. The methods of blockading it will obviously be a combination of mechanical and vegetative obstructions, and with these foresters should be particularly qualified to deal. This fact further accentuates the responsibility of the Forest Service, and indicates that the watershed work of the future belongs quite as much to the forester as to the hydrographer and engineer.

The Fuels Buildup in American Forests: A Plan of Action and Research

1971. *Journal of Forestry*. Vol 69. No. 8. 471–475.

Carl C. Wilson and John D. Dell

Today more than a billion acres of forest and rangeland in the United States are managed under some form of organized fire protection. On much of this wildland, there is a buildup of flammable fuels that under critical burning conditions can feed disastrous forest fires. The continuing trend toward intensive forestry will, in the long run, contribute to reduction of this wildfire potential, but the problem remains a serious one.

Fire protection in the United States has come a long way in the twentieth century. But many foresters have convinced the public, and even themselves, that mechanization and armies of trained forest firefighters are sufficient to handle any threat from fire in our forests. Unfortunately, it is not, as Arnold (3) and others have shown.

Throughout he nation, hills and wildlands have become suburbs of growing cities. Thousands of homes have been built in critical fire areas, and more can be expected. Fuel buildup in these areas must be controlled or a continuous cycle of fire, devastation, regrowth, and fire will occur.

In California during 1970, nearly one-half million acres were charred as dozens of fires, blasted by Santa Ana winds, spread wildly through forests, brushlands, rural communities, and urban residential areas. The 175,000-acre Laguna fire in San Diego County was the second largest single fire ever recorded in the state's history (*Fig. 1*).

Just nine years earlier, in 1961, the Bel Air holocaust in Los Angeles blackened more than 6,000 acres of valuable watershed and burned 505 buildings and residences valued at $30 million. Such fires take a tragic toll in human lives as well. Twenty firefighters were killed on the Loop and Canyon fires in Los Angeles County in 1966 and 1968. In this land of dry brush, of steep canyons and sun, and of houses perched on hillsides of chaparral, some of the best trained firemen in the world, along with the most modern aerial and ground equipment available, cannot prevent these costly disasters where

The authors are, respectively, Carl C. Wilson (left) assistant director, Forest Fire Research, Pacific SW Forest and Range Exp. Sta., Berkeley, Calif.; and John D. Dell (right) research forester, Forest Protection Research, Pacific NW Forest and Range Exp. Sta., U.S. Forest Serv., Portland, Ore.

dense, continuous fuels and critical weather conditions are so much a part of the fire environment.

Large fires, however, are not limited to California alone. In July and August 1970, over 200,000 acres of prime timber and watershed were destroyed on national forest and state-protected land in central Washington. In 1962, fires in New Jersey caused the death of seven persons and burned more than 195,000 acres—destroying 500 homes. A single fire burned 60,000 acres in one day.

Idaho's Sundance fire, September 1, 1967, made a spectacular run of 16 miles in nine hours, claimed two lives, threatened several communities, and destroyed more than 50,000 acres of valuable timber and watershed. Anderson (1) estimated that at one point the fire was releasing thermal energy of nearly 500 million BTU per second, which is equivalent to a 20-kiloton nuclear bomb exploding every 2 minutes. A mobilized force of several thousand men combined with favorable weather were needed to stop this fire.

These examples only emphasize the futility of setting mechanized man against the destructive forces of wildfire where fuels have accumulated.

What can be done to reduce the number of destructive forest fires? We know that conditions of atmosphere and topography can fortify or weaken the driving force supplied by fuel. Unfortunately, our knowledge is not sufficient at present to modify fire weather conditions or to appreciably reshape the topography. We can, however, manipulate and modify the forest fuels. The job cannot be done overnight, and the solutions may not come easily. However, the investment in the protection and enhancement of our environment will be invaluable to us and to those who inherit the land from us.

This paper describes some of the more conspicuous problems of fuel buildup, evaluates what can be done about them with our present knowledge, and outlines high-priority research needs.

Nature's Contribution to the Fuels Problem

The natural cyclical process of growth and mortality the source of major fuel buildup. All of the living is and functioning parts of trees are renewed continuously, and their dead parts may accumulate anywhere from forest floor to canopy top. Shrubs, grasses, and herbs may also contribute to this fuels complex.

Figure 1. Road construction slash has high fire hazard potential and is a poor advertisement for good forest management.

Nature contributes to the fuels problem in other ways. Natural disasters, such as droughts, windstorms, insect epidemics, and disease infestations, increase vegetative mortality. When the proportion of dead material in the fuel structure increases, there is an increase in the rate of fire spread, bum intensity, and spotting potential.

Man's Contribution to the Problem

It has been amply demonstrated that the forest manager today must be concerned with the quality of the forest environment. The impact of some of our existing forest practices on the environment must be considered in future management decisions. If we wish to grow and harvest forest vegetation, we must be prepared to manage the fuel situations we create. Many of our silvicultural practices leave behind them an intolerable quantity of debris. The immediate effect of timber harvesting, thinning, and forest road construction may be accumulations of smashed logs, limbs, tops, rotten wood and broken-down underbrush.

What is the duration of the hazard of this man-made fuel? Fahnestock and Dieterich

(8) have found that foliage and branchwood slash i n Idaho lost virtually all its needles in five years. Meanwhile, rot appeared in the branchwood of all species. Although the hazard was less, sufficient fuel remained for future fire potential. A long-term study, begun in 1927, on California's Lassen National Forest, showed that medium-to-large branches, logs, and stumps were still intact 30 years later.

Debris from logging, thinning, or forest road construction becomes susceptible to ignition as it decomposes and after many years may be as easy to ignite as dry grass. Slash burning by itself is often not enough to eliminate the fuel problem completely. Autumn bums may dispose of the flash fuels but only scorch large cull logs and other heavy fuels. This material eventually becomes available fuel again.

Logging, thinning, and road construction open up the rest and increase the amount of sunlight and wind at ground level. Countryman (4) has estimated that opening up a virgin, mixed-conifer stand could increase rate of fire spread up to 4.5 times. This estimate considers only the changes in microclimate—not the dead fuels created.

Morris (11) reported that, assuming other conditions remain constant, the fire weather in clearcut Douglas-fir slash in the late afternoon will be seven times more severe than in the adjacent, uncut, timber.

In a study of the fire hazard from precommercial ponderosa pine thinning slash on the Deschutes National Forest in Oregon, Fahnestock (7) estimated that nearly 40 tons of dead fuel per acre exist on some sites thinned to an 18-by-18-foot spacing. The question of whether fires can be more easily controlled in thinning slash or in dense thickets of reproduction is debatable. Nevertheless, thinning slash, like other man-made debris, provides a flammable fuelbed in which fire can ignite and spread.

Construction debris along permanent and spur logging roads can be a fuel hazard if left untreated (*Fig. 1*). Dell (5) reports that roadside slash on interlinking roads between freshly logged clearcuts became connecting fuses for fire spread on the 1967 Raft River fire in Washington.

Another hazard is created by even-aged "brushfield" conifer plantations—especially where dense stands override old logging slash. There are an estimated 20 million acres of plantations in the United States. Each year fire wipes out thousands of acres of these tightly spaced, continuous young stands. In May 1965, 11,000 acres of plantation on the Nebraska National Forest were consumed in a devastating wildfire. More recently, in 1968, all three of Michigan's largest fires during the spring fire season involved plantations. In most cases, the fires started in grass outside the plantations and spread inside. Although even-aged plantations are easier to establish initially, researchers are concerned that such plantations may be more susceptible to insect and disease attack than uneven-aged stands. The fuel hazard they represent reinforces the arguments against the practice.

What Can Be Done Now?

What, then, can we do about the existing fuels problems in our forests? To begin with, fuel treatment planning must *precede* activities that modify or change vegetative conditions. Hazard reduction programs in *natural* fuels must also be implemented—with priorities given to the most critical areas at first.

The job of reducing fuels and treating debris from forestry operations must deal with far more than fire control alone. Although the fire specialist is usually responsible for fuels management, he must consult, plan, and work directly with timber, engineering, water shed, recreation, and range and wildlife managers. He should also work with the landscape architect and those responsible for air and water pollution control to insure that effective fuels management is in harmony with environmental objectives.

Prescribed burning.—Fire has always been an integral part of the forest environment in North America. It is the natural agent for thinning, pruning, and fuel reduction. Man's good intentions in protecting the forest from fire are indirectly adding to the fuels problem.

The exclusion of fire from the natural environment may also be causing some ecological imbalances. Hartesveldt (9) reports that without fire, the giant sequoia of California could not have evolved or survived. There is a question as to how long it can continue to survive under the intense protection it now receives. This is not to imply that fire is always beneficial to the forest. But, when properly controlled and scientifically applied, fire can be used to the advantage of man for the improvement of his environment.

Some foresters say that prescribed fire use has a very limited future because the public will not tolerate wood smoke in the air. Indeed, if smoke cannot be managed effectively, this important tool for fuels treatment may indeed be lost. In some areas, location of prospective fuel treatment areas close to population centers (perhaps on all sides) may preclude any type of open burning.

On the other hand, carefully prescribed burning can be accomplished near smoke-sensitive areas if a combination of conditions—smoke source, elevation, fuel moisture, wind direction, and atmospheric stability—are favorable for effective smoke dispersion. This was illustrated in 1969 in Oregon, when during a period of stable air conditions in the Willamette Valley, more than 6,000 acres of widely located slash units were burned at higher elevations in the Cascade Mountains. Slash smoke dispersed eastward on favorable winds aloft, and did not influence the locally polluted air of the valley (6).

By applying sound meteorological principles to smoke management, the forester can still use prescribed fire scientifically to improve the environment without polluting it. Here, again, forest management must be geared to contemporary needs.

Fuel treatment.—The boundaries between urban and rural lands are becoming less definite every year. Land use must be predicted and integrated in fire control planning. A major step in this direction is the fuel break program, whereby brush or timber complexes are broken up by wide strips or blocks of land on which vegetation has been permanently modified to directly aid fire control. Although this program exists to some extent in all parts of the country, it needs greater acceleration and financing—both for construction of new fuel breaks and maintenance of old ones. Fire control should be made part of the total resource planning job. Fuel breaks and other vegetation modification projects can be constructed ahead of, or in conjunction with, timber sales and thinning projects, or the layout of large conifer plantations.

In high-value watershed areas (such as southern California) where water for fire control is scarce, fuel break planning should incorporate strategic placement of water storage and distribution facilities. Also, in these critical fire areas where communities and flammable forest fuels intermingle, local officials and fire control agencies should consider utilizing reclaimed waste water for irrigating "greenbelts" or safety zones.

Fuel modification over vast areas may be an eventual requirement for effective fuels management. Meanwhile, however, we need to make a greater effort to apply our knowledge of prescribed fire for creating and maintaining fuel breaks—especially in timber areas where fire can be used to remove dead fuels and highly flammable vegetation in the understory. Low intensity control burns can contribute to a parklike forest and improve stand condition.

In logging slash areas, the forester should plan more complete cleanup of larger fuels (Fig. 2). Where topography is suitable, tractor piling or windrowing of slash should be considered in lieu of broadcast burning. Piling and burning leaves a logged area tidier, and allows compacted fuels to be burned at more convenient periods for effective smoke management. Even on steep slopes, large material can be cable yarded and piled at landings or other locations. These piles, if well concentrated, will burn even after long wet periods.

Utilization of forest residue.— Removing some waste material from the greatly reduces fire hazard. Better utilization of forest residues depends largely on market conditions and needs for the raw material. In the West, the most significant use for residues is, at the present, for pulpwood for domestic use and export. Nearly one billion cubic feet of pulpwood is needed annually. A portion of this requirement can be met by chippable culls left after logging. However, on national forests, the prospects

Figure 2. Clean logging and concentration piling of cable-yarded fuels for later burning leaves cutover areas with a neat, well managed appearance.

for removal and utilization of such material now are limited by high costs of handling waste materials.

Utilization of past research.—There is a need for foresters to make better use of the vast amount of research information already available to them. The communication gap between the researcher and the forester must be bridged.

More than 10 years ago, Arnold (2) suggested that very fire control organization of any size should have one permanent position assigned to the nearest appropriate fire research group. In critical problem areas such as fuels management, the liaison man should work with the research organization on applied research projects and communicate new findings directly to field users. In turn, he should inform research of the most critical field problems.

We have made good use of highly professional educational and training programs in recent years to develop skilled fire specialists. The U.S. Forest Service's National Fire Training Center in Arizona conducts schools in fire behavior, prevention, generalship, and law enforcement. A national fuels management course could also be instituted to communicate new concepts and techniques.

Communication with the public.— Finally, there is an urgent need for foresters to communicate more effectively with the general public on the importance of intensive fuels management. They need to explain and demonstrate their programs and show the results of their efforts.

Too often in the past, foresters have become defensive when their methods or objectives were questioned. They wince when they read headlines about slash burning polluting local airsheds—although they may know the cause was actually industrial pollutants. How much more effective to take the dynamic approach and explain to the public their smoke management policies, research efforts, and the results of these efforts directed at improving the environment.

Such public relations programs should begin well in advance of fuel treatment projects—not after they are completed. The public must be made aware of the conflagration potential of forest fuels buildup, the need for treatment measures to reduce or eliminate this hazard, and the role of prescribed fire as an ecological tool.

Research Needed

Progress in fire control over the years has resulted om a combined effort by research and management. The fire research job is a long way from completion with new and challenging frontiers ahead. There must be more research emphasis on forest fuels management as a part of the environmental forestry required today and in the future.

Fuel classification.—First, researchers must develop a functional nationwide fuel evaluation, inventory, and appraisal system. This will allow more objective planning of effective control programs. The system should be tied in to the new National Fire Danger Rating System so that manning and action guides could be considered for specifically classified fuel areas.

Also much needs to be learned about the use of prescribed fire under natural conditions. We particularly need to know more about the interactions of fire intensity on seedling growth in different forest tree species. Some Australian foresters have determined that, in certain species, a high intensity fire produces an "ash bed effect" that promotes good tree growth. In the state of Victoria, good results have also been achieved with controlled low intensity bums to reduce the potential for the large conflagration. Hodgson (*10*) reports that more than 500,000 acres are control burned annually in Victoria, ,and that this contributes to a marked reduction in total wildfire occurrence and size of areas burned.

In the United States, similar studies are needed in critical areas like California's brushfields to determine the range of weather and fuel conditions under which prescribed burning for hazard reduction can be accomplished safely and effectively. Aerial application of desiccants and smashing for fuel preparation need further exploration. These may permit burning of brushfields at times when surrounding green vegetation is high in moisture content with low spread and spotting potential outside the prescription area.

Our concern with forest fuels management should not be limited to "front country" problems only. In the national forests and national parks, public use of wilderness land is increasing rapidly. In Oregon and Washington, for example, there are nearly 2 million acres of national forest wilderness areas. In 1959, these areas received 145,000 visitor-days of use. In 1969, however, the total visitor use was 700,000—an increase of nearly 400 percent in 10 years.

We have not even begun to recognize what the role of re may be in wilderness area management. Fire, once the natural agent for forest fuel reduction, has been almost eliminated by protection. Natural fuels continue to accumulate, and—with heavy public use, very limited access for vehicles, and typical steep, rough topography—the potential for large fires increases every year. Fire research and ecological study into the con trolled use of fire in these high-value areas are necessary. One example of such research is the new study by the National Park Service in the Sierra Nevada of California, centering on the use of prescribed fire in the true fir type and its ecological effects on ground litter, stand density, brush and tree seedling development, and, of course, fire hazard abatement. Other prescribed fire studies are being conducted in other forest types—including the giant sequoia.

Hazard modification studies.— Establishment of fuel breaks and other vegetation conversion projects is use less unless these areas are constantly maintained. Research must continue to investigate and develop economical and efficient chemical, mechanical, or physical methods for doing this. More study must be given to strategic locations for fuel breaks to protect communities and residential areas in and near hazardous fuel areas.

The search for "low volume" and "slow burning" vegetation as a replacement for existing hazardous fuels is important in fuels management—especially for critical brush areas.

Another hazard reduction technique needing more intensive study is chemical fire-proofing for roadsides and railroad rights-of-way. In locations where the critical fire season is short, or where large-scale fuel modification is not feasible, fireproofing with nontoxic, non persistent chemicals could reduce or eliminate ignition problems on access routes that have a high incidence of fire. Such a chemical should be water resistant—with at least the capability of lasting through normal fire season precipitation.

Engineering and equipment research.— Forest fuel modification and debris removal will require development of new and specialized equipment for specific jobs.

Equipment capable of modifying forest fuels is needed to (a) establish fuel breaks; (b) reduce fuels to smaller components to hasten decomposition; (c) compact the fuels to increase chances for decay; and (d) improve access through the fuels complex—all without damage to the forest resources.

For utilization of forest residue, equipment must be developed that can efficiently and economically extract, transport, and load large or small material from steep slopes and over long distances. In areas remote from processing centers, large portable systems need to be designed to modify the residues to a more transportable form right at the woods site.

Economics research.—Foresters can change management policies and amend timber sale contracts toward better logging residue utilization, but research must develop sound economic reasons for the decisions made by management. Cost-benefit ratios for better utilization must be determined both for landowners and logging or land clearing contractors.

Research needs to devise accurate and practical techniques for measuring residue volumes after timber harvesting, and should develop methods for equitable pricing of salvageable material. More study is needed of the true importance of good forest fuel modification not only in fire control, but also in the many concerns of forest management for ecological, commercial, and human values.

Literature Cited

1. Anderson, Hal E. 1968. Sundance fire: an analysis of fire phenomena. U.S. Forest Serv. Res. Paper INT-56. 39 p.

2. Arnold, R.K. 1959. Where we stand In forest fire research. Proc., Western Forest Fire Research Council Meeting, Spokane, Wash.: 2–5.

3. ———. 1967. The forest conflagration consequences of management decisions Proc., 14th IUFRO Congress, Munich, Germany: 782–788.

4. Countryman, Clive M. 1955. Old growth conversion also converts fire climate. Proc., Soc. Amer. Foresters Annual Meeting, Portland, Ore.: 158–160.

5. Dell, John D. 1970. Road construction slash-potential fuel for wildfire. Fire Control Notes 31(1):3.

6. Dell, John D., F.R. Ward, and R.E. Lynott. 1970. Slash smoke dispersal over western Oregon . . . a case study.

USDA Forest Serv. Res. Paper PSW-67. 9 p.

7. Fahnestock, George R. 1968. Fire hazard from precommercial thinning of ponderosa pine. U.S. Forest Serv. Res. Paper PNW-57. 16 p.

8. Fahnestock, George R., and J.H. Dietrich. 1962. Logging slash flammability after 5 years. U.S. Forest Serv., Intermountain Forest and Range Exp. Sta., Res. Paper 70. 15 p.

9. Hartsveldt, R.J., and H.T. Harvey. 1967. The fire ecology of sequoia regeneration. Proc., Tall Timbers Fire Ecology Conf. 7:65–77.

10. Hodgson, Athol. 1968. Control burning In eucalypt forests In Victoria, Australia. J. Forestry 66(8) :601-605.

11. Morris, William G. 1941. Fire weather on clearcut, partly cut, and virgin timber areas at Westfir, Oregon. Timberman 42(10) :5–8.

Problems and Priorities for Forest Fire Research

1973. *Journal of Forestry*. Vol 71. No 10. 625–628.

 Craig C. Chandler and Charles F. Robert

Forestry is changing in both orientation and focus. Not since the opening years of the 1930's has the profession faced the excitement, turmoil, and dissatisfaction that accompany a restructuring of values. The direction, the magnitude, and the pattern of these changes are of fundamental importance to the identification of future research needs.

It now seems certain that future historians will mark the middle 1960's and the early 1970's as the years when the wave of environmental awareness broke suddenly and significantly on the whole of human society. Forestry is near the center of many key environmental issues, and the forester is now being forced as never before to subject his management and protection policies to the harsh light of public scrutiny.

One result is a recognition of two criticisms leveled at research (with some justification, we believe): One, that there is too much concern with abstract problems of interest to the researcher but of little value to the practicing forester; two, that when a practical problem is attacked, the research and development of solutions often outlast the actual problem. If research is to help

solve forestry problems of the coming decade, it must respond to the practitioner's needs, must produce applicable results in a shorter time, yet must not lose its objectivity and scientific quality

Another result is the recognition that forestry does not always have within its repertoire of planning and management techniques exactly those needed to achieve a specific set of goals. Hence many foresters have become increasingly uncomfortable with current practices, intuitively recognizing that many practices do not always accomplish the stated objectives, or that some objectives are not really responsive to changing needs and goals. In protection from fire, these nagging doubts and uncertainties have been particularly persistent.

Reevaluating Protection Policy

Early in the history of organized fire protection, decision makers recognized that this science, like most other forms of human endeavor, is subject to its own form of diminishing returns. Once a modest level of protection capability has been attained, additional increments of suppression effectiveness are purchased at a

The Authors—Craig C. Chandler is director; Charles F. Roberts is meteorologist, Division of Forest Fire and Atmospheric Sciences Research, Forest Serv. U.S. Dep. Agr., Washington, D.C.

rapidly increasing unit cost. In the face of an enormously variable fire hazard and risk, it is difficult to decide just what constitutes adequate protection. After 50 years of sporadic research and policy formulation, the question remains intractable.

Two new factors have recently contributed to the complexity of this policy problem. First, a growing recognition of some of the beneficial effects of fire and its increasing use as a forest management tool. For example, the area burned by prescribed fire in 1970 has been estimated at 2.5 million acres, while wildfires burned only 2.25 million acres. Because these prescribed burning operations were carried out at considerable expense, we can conclude that fire in the forest can have economic as well as ecological benefits.

The second factor has been the skyrocketing costs of fire suppression. In the past decade annual costs have increased from $65 million to $117 million. After the disastrous fires of 1970 in Washington, Oregon, and California, the question of justifiable

protection costs triggered a thorough review by the chief of the Forest Service, his deputies, and regional foresters (3). This review failed to arrive at an acceptable alternative to the so-called "10 a.m. policy" (requiring an all-out attack on any and all fires so as to bring them under control in the next 24 hours), but it led to a series of continuing evaluations which are now producing major changes in the outlook for fire protection.

It is, perhaps, not too simplified to say that these valuations have resulted in a revised set of working assumptions to guide the development of fire-protection programs of the future.

- Fire in the forest is neither good nor bad, *per se*, except in terms of the degree to which it promotes or hinders land management objectives.

- Not all wildfires are susceptible to control by methods which conserve a proper cost-benefit relationship.

- The objectives of fire protection can often be effectively promoted and to some extent be even totally achieved through the judicious planning and execution of resource production, management, and harvesting activities.

- There is no single "best" solution or panacea for wildfire problems. The solution will depend on a complex interplay involving values at risk, land use and management objectives, burning conditions, and suppression force capabilities.

These assumptions are not only working axioms; they also identify the most important fire research needs for the next five to 10 years:

1. New methods and concepts of hazard or flammability reduction.
2. Accurate prediction methods for fire behavior and the specification of fire effects.
3. Improved risk specification and risk reduction methods.
4. Adaption of new technology for planning and executing protection actions.

Hazard Reduction

It appears that the highest priority for fire research, both in potential payoff and in probability of success, lies in hazard reduction through fuel management. Two traditional approaches to hazard reduction—fuel breaks and residue reduction-have been under limited research for years. Establishment and maintenance of fuel breaks are now nearly state-of-the-art, but present methods are too expensive for widespread, routine application, and there remains uncertainty about their effectiveness. Nor is it clear that the present approach to fuel break creation is susceptible to significant cost reduction or improved effectiveness of fire protection. Consequently, other approaches to hazard reduction should be sought which might

reduce capital investments and maintenance costs. There are no obvious answers. Hence, we assign highest priority to a reanalysis of management of naturally-occurring fuels.

Forest residue-reduction programs are also important. The Forest Service recently started several intensive efforts toward improvement of technology and application. Although there is still far to go before this effort meets the scope of the problem, three general methods of residue reduction are under investigation: full utilization of logging residues and debris; mechanical or manual removal of understory vegetation and fuel; and prescribed burning under a wide variety of fuel situations and timber types.

Full-utilization studies will test several new options, the most exciting being the complete use of logging residues as fuel for small steam-driven electric genera tor plants. If this proves feasible, not only will it reduce a major source of unwanted fire, it will also help alleviate the nation's energy shortage. Other uses involving the manufacture of wood products are equally appealing and more immediately promising; never the less, it is interesting that in 1970 wood supplied a higher proportion of our energy needs than nuclear fuels—900 trillion BTU from wood (6) and 80 trillion BTU from nuclear fuels (excluding naval transport) (1).

For the mechanical removal of uneconomic residues, no ideas worthy of major R&D effort have yet been put forth, and prospects for the discovery of reasonably practical cost-effective methods do not appear bright.

Research on prescribed burning for residue reduction must cope with smoke management, ignition and control techniques, and the development of relations which describe fire effects in terms of measurable burning conditions. Because of natural differences in cover types, topography, and climate, operational use of widescale prescribed burning will continue to come into practice in fits and starts. But there is no doubt that fire will be more widely used as our ability to write safer and more definitive prescriptions improves.

Fire Behavior and Effects

Fire occurrence and damage statistics have always supported the finding that most fire losses stem from a very small fraction of a season's fires (the figures are generally arrayed to show that less than 5 percent of the fires produced more than 95 percent of the damages).

To achieve the goal of minimum social cost plus loss from fire requires time-dependent decisions on resource allocation, in which the key variables are fire behavior, fire effects, and resource values. Despite long concern, very little quantitative information has yet been compiled on the fundamental relation between fire behavior characteristics and effect on forest resources. That is, there is no known measurement or calculation which can specify, *a priori,* fire damages under a given wildland fire situation.

Recent efforts by Rothermel (7) and Frandsen (5) have produced one-dimensional, linear steady-state models of fire spread and energy release suitable for calculation by electronic computers. These models were developed from empirical data (laboratory fires) and physical principles, but they have not yet been fully validated with field measurements. This defect needs early rectification.

Ultimately, the need for predictions of fire behavior and effects will require development of generalized, three-dimensional growth models for wildland fires. The great variability which characterizes the phenomena involved in fire behavior will likely require stochastic rather than deterministic descriptions, especially since the weather situation appears to be a decisive element

in the growth and decay of fire spread and intensity, and the resultant impacts on forest resources. The lack of predictability in some of the weather elements, especially windflow near the fire, has devastating implications for deterministic fire-prediction and fire-damage models.

Prediction of fire effects will require not only good specifiers of fire behavior and characteristics, but also valid relationships between these characteristics and the economic as well as physical and biological effects on various components of the forest resource. Fire research must place increased emphasis on research and analysis involving the use of prescribed fires, and careful documentations of wildfire behavior and subsequent physical alteration of the ecosystem.

Risk Reduction

Obviously, reduction of risk would have a dramatic effect on fire occurrence rates. The problem is that we cannot visualize any really effective practical methods to reduce risk significantly. For our purposes, we define risk as the probability of a firebrand with fire-starting potential coming in contact with a wildland fuel complex.

Traditionally, firebrand sources have been classified as man-produced or lightning-produced. Recent studies have found it important to further classify man-caused risk into deliberate and accidental risk, and stratify deliberately induced risk into that resulting from rational and irrational acts (2).

Clearly, these classifications cover a wide spectrum of human activities; this, as much as any other element, accounts for the great difficulty in developing effective risk-reduction methods. In effect, the problem amounts to efforts to alter human behavior and habits in a large percentage of day-to-day activities.

The general nature of risk (both man-caused and lightning) suggests that reduction methods must be scheduled on the basis of temporal and spatial variations in risk. Hence, we would assign highest priority in fire-prevention research to the development of reliable risk specifications and then to risk reduction methods.

The most tractable aspect of the risk problem seems to be that of lightning-produced risk. For the past 10 years, lightning has received more research effort than any other fire-causing agent. As a result of work on lightning abatement by the Forest Service Project Skyfire, technology will soon enable us to establish the costs, benefits, and guidelines for operational application of cloud seeding.

Adapting New Technology

Earlier, we pointed out that one of the major impacts of the new look in forestry will be the increasing need for a workable, comprehensive planning method, permitting treatment of the forest resources as a single integrated system. This requires that the components of the forest environment— soil, water, air, trees, wildlife, shrubs and grass—be developed and managed synergistically, not as separate entities. All plans for production, management, or harvest of these resources must be part of the same over-all management objective. Clearly, the protection programs must be planned, designed, and carried out within the same framework.

Planning such highly interactive programs is an enormously complex task. Fortunately, new techniques form the analysis and design of complex feedback systems are being worked out for uses other than development of hardware and instrumentation (see (4) for example). These methods must be adapted to planning and policy evaluation for forest resources management.

Effective planning must always be based on valid data; thus, a vastly improved data bank on fire occurrence, fire damage, and total resources is an urgent requirement. The methods for acquiring, storing, and processing these data are in part a research responsibility, and fire research in the future ought to contribute to this important activity.

To upgrade the capability to perform a specific task without significant cost increase, improvements in both efficiency and effectiveness are required. In nearly all forms of human activity, this has been achieved through the introduction of new technology. Agricultural production provides a striking example of how the output from a fixed or declining input (labor force and land area under cultivation) can be increased spectacularly by introducing new technologies. We believe that a similar opportunity exists in some parts of the fire suppression operation.

Two facets have already benefited from adaptation of new technologies: detection and aerial attack. Research to use remote sensing methods in fire detection has resulted in an operational IR detection and mapping system mounted on a high-performance aircraft (8). In the next few years, we envision dramatic new applications of remote sensing for fire protection, especially developments in laser spectroscopy and forward looking infrared. If successful, these adaptations will result in a major improvement in fire detection and initial attack capabilities.

The use of computers for real time monitoring, analysis, and prediction is now commonplace in many military and industrial operations. This capability should and can be extended into fire suppression operations, especially as an aid in selecting strategy and tactics, in resource inventory and scheduling, as well as in general fire supp ion decision making. The new concepts of a mobile national fire attack and suppression force that are currently being considered would be greatly enhanced ·by providing this type of support capability.

To sum up, we believe that forestry faces a change some of the fundamental doctrines that have shaped in management and .protection practices in the past. As a result, forestry will be assigned as its principal responsibility the proper planning and management of the nation's wildland environment under new objectives promulgated by an aroused and, we hope, informed citizenry. These changes result in a new set of axioms for formulation of fire-protection policy and for the identification of fire research needs. The new look in fire management will increase the urgency of many research, requirements. Some involve new problems, but most will result from shifts in emphasis and priority of problems that have been around almost as long as organized fire protection. To improve output, measured in terms of mission accomplishment, without major increases in required input, fire research will need to keep abreast of the developing technology in all of the applied sciences and identify those advances which can be usefully adapted to fire protection needs of the future.

Literature Cited

1. Atomic Energy Commission. 1972. Operating history of U.S. nuclear power reactors.

2. Doolittle L. M. 1972. The dimensions of man-caused forest fire risk: a systematic assessment. Ph.D. Dissertation. Univ. Washington, Seattle.

3. Forest Service, U.S. Dep. of Agr. 1971. Fire Policy Meeting, Denver, Colorado, May 12–14, 1971.

4. Forrester J. W. 1968. Principles of systems. Wright Allen Press, Cambridge, Mass.

5. Frandsen W. H. 1971. Fire spread through porous fuels from the conservation of energy. Combustion and Flame 16:9–16.

6. Hottel, H. C., and J. B. Howard. 1972. An agenda for energy. Technology Review. Mass. Inst. Techn January 1972. p. 39.

7. Rothermel, R. C. 1972. A mathematical model for predicting fire spread In wildland fuels. USDA Forest Serv. Interm. Forest and Range Exp Sta. Res. Paper INT-115.

8. Wilson R. A., S. N. Hirsh, F. H. Madden, and B. J. Lorensky. 1971. Airborne infrared forest fire detection system: final report. USDA Forest Service Interm. Forest and Range Exp. Sta Res. Paper INT-93.

Managing Fire in Our Forests—Time for a New Initiative

Stephen F. Arno and James K. Brown

Stephen F. Arno is research forester and James K. Brown is project leader, Fire Effects and Prescribed Fire Research Unit, Intermountain Research Station, USDA Forest Service, Missoula, MT.

An impressive body of scientific evidence on fire history, fuel accumulation, and fire behavior makes it clear that much of North America is a "fire environment" where wildfire or a substitute recycling mechanism is inevitable (Anderson and Brown 1988, Arno and Wakimoto 1988, Wright and Bailey 1982). When fire is suppressed for long periods, epidemics of bark beetles, defoliating insects, and diseases often allow heavy accumulation of fuel—a prime requisite for catastrophic fires.

In the late 1970s the USDA Forest Service greatly expanded its fire control policies to encompass "fire management" (Nelson 1979), which includes using prescribed fire (ignited by land managers or lightning) for land management objectives such as decreasing hazards, improving wildlife habitat, and maintaining wilderness ecosystems. Today, prescribed burning is an accepted practice in many American forests. It is commonly practiced in the South and parts of the West for control of undesirable species, hazard reduction, and site preparation.

Despite widespread adoption of there management concept a decade ago, forest fuels continue to increase faster than they are being recycled through harvesting, fire, and decomposition. Concerns about excessive smoke, the possibility of fire escapes, large fuel accumulations, steep terrain, and difficult access have restricted the use of prescribed fire. Vegetation and fuels management with prescribed fire or other cultural methods has also been hampered by lack of public awareness of the fuels build-up problem and by shortages of funding and personnel. In contrast, vast sums of money are spent attempting to control severe wildfires in untreated fuels.

The past few years have produced dramatic testimony to the folly of ignoring the buildup of wildland fuels. For example, more than 1,400 homes were lost to wildfires nationwide in 1985, with Florida and North Carolina suffering the most damage. In 1987 dozens of homes in northern California and southwestern Oregon were destroyed, and thousands of others had to be protected at great cost. The situation was repeated even more dramatically in 1988 in the greater Yellowstone region, the Black Hills of South Dakota, and across Montana and northern California. Dozens of homes were destroyed. Thousands of homes and several multimillion-dollar resorts had to be protected at staggering expense.

Most fire suppression experts believe that the threat of massive damage to human lives, private property, and natural resources is increasing (National Fire Protection Association 1987, Fischer and Arno 1988). In 1988, $145 million was spent to suppress the Yellowstone-area fires—largely to combat threats to homes, resorts, and facilities. Another $4 million was spent battling the intense Red Bench Fire, in and adjacent to Glacier National Park, which burned through heavy beetle killed forest fuels and destroyed twenty-one dwellings. Large sums must now be spent to rehabilitate many miles of dozer-built fire lines. In the same year, the Canyon Fire escaped from the 1.5-million-acre Bob Marshall Wilderness complex (MT) and raced across 50,000 acres of private land to-ward Augusta. Several uncontrollable wildfires also burst out of the Selway-Bitterroot Wilderness and threatened to overrun woodland homes in Montana's Bitterroot valley. Despite massive suppression efforts, these 1988 fires were uncontrollable until the weather turned. The lone exception was the Little Rock Creek Fire, which was controlled where it encountered thinned stands (with light fuels) and an earlier 1988 burn.

When homes and developments lie in the path of a wildfire, firefighting efforts must be diverted to saving structures—which in turn allows the fire to grow and threaten more homes and forests and this is a growing nightmare for firefighters and land managers.

A Fuels Management Program

In addition to high suppression costs, severe fires are extremely costly to lives, property, and natural resources. What, if anything, can foresters and re source managers do to reverse the dangerous fuel accumulation that makes such fires inevitable? Forest biologists and fuels specialists, particularly in the West, believe one answer is a broader application of prescribed fire and silvicultural cutting substitutes for fire. This article proposes three different strategies that can be applied to three types of zones: wilderness and natural areas, the general forest management zone, and the residential forest.

Zone I—wilderness and natural areas. In large wilderness and natural areas, "prescribed natural fires" (lightning fires allowed to burn under previously approved conditions of weather, fuel moisture, time of year, etc.) can reduce fuels and maintain a vegetative mosaic of old and young stands. Ideally, such management would create a forested area in which fire would have the same role as in primeval days. However, events in the 1970s and 1980s suggest that this is an unrealistic expectation, at least in the foreseeable future. Many lightning fires must be extinguished because they threaten lands outside the natural area. As illustrated in 1988, early and mid-season fires and those originating near area boundaries present a risk to resources and facilities outside the wilderness. These risks are exacerbated by half a century of fire exclusion, which allowed living and dead fuels to accumulate over a wide area. The original role of fire is also circumvented when lightning fires in adjacent lands must be suppressed to protect settlements, agriculture, timber, and other values. Thus the flames from adjacent areas no longer spread into wilderness or natural areas.

Foresters and fuels managers need promote a more realistic view of the to alternatives-such as manager-ignited prescribed fires-for reintroducing fire in wilderness and natural areas (Reeves 1989). Continuing to exclude fire from most small natural areas and wildernesses circumvents the basic ecological processes and eventually can lead to severe wildfires.

Zone II—general forest management zone. The more accessible forests in this zone are abundant and provide a transition between Zone I and Zone III. When timber harvesting is properly carried out, good fuels management can be a result. Timber harvesting may reduce biomass sufficiently without prescribed fire. However, in many areas dense young second-growth thickets (living fuels) grow up after harvesting. Also, stands with marginal economic value continue to accumulate fuels. In stands where fuels need to be reduced, visual constraints sometimes discourage harvesting even though a severe wildfire could be devastating visually. Trees are not immortal; change in the forest is inevitable, but it can be guided through management.

Although fuels in this zone can be managed well, such management is often neglected because of costs, administrative constraints, and lack of incentives. The public must realize that we are jeopardizing our forest resources and ignoring ecological processes by shortcutting fuels management.

Zone III—residential forest. Hazardous forest fuels in Zone III surround or border homes, summer cabins, resorts, and communities. Dead and living fuels normally accumulate in these forests, but most citizens are unaware of the dynamic growth and death processes in wildland ecosystems. Structures within this zone often com pound the problem by being highly flammable themselves-with cedar shingle roofs, landscaping bark, and stacks of firewood nearby. Manipulative management in commercial forests within this zone can effectively reduce fuel hazards. In small private owner ships, however, such management is possible but unlikely.

When fires threaten Zone III areas, control is compromised because the suppression resources and priorities are committed to protecting individual homes and other developments. In the 1980s, numerous examples from Florida to Michigan and from California to Washington illustrate that widespread drought, and an accumulation of forest and shrubland fuels, fostered large uncontrollable fires that threatened residential developments. This greatly complicated suppression effectiveness (National Fire Protection Association 1987, Fischer and Arno 1988). Fires started by lightning, human carelessness, or arson often escaped vigorous initial attack. Even ignition prevention and rapid response may not prevent severe damage to wildland resources.

Homeowners and community administrators must understand that growth and aging of unattended vegetation gives rise to hazardous fuels. Flammability can be reduced by manipulating fuels in esthetically pleasing ways. Homeowners who allow hazardous fuels to accumulate on their property should be aware of the risks of uncontrollable fire and be willing and able to assume the costs of damage.

A New Initiative

Under present management, wildfires have reached crisis proportions in one or more regions during most of the 1980s and could easily get worse. Technical knowledge is already available to substantially reduce the threat of wild fire, but it cannot be implemented without public support. Foresters and fire specialists need to convey to the public the seriousness and immediacy of the fuels problem and the availability of solutions (Reeves 1989). The ultimate goal is to reduce the threat of wildfire damage to people, homes, facilities, and natural resources. Efforts should be concentrated in Zone III and adjacent portions of Zone II.

Simple, intensive forestry techniques could vastly reduce the probability of wildfire damage by reducing potential fire intensity and rate of spread. Fire behavior modeling (Anderson and Brown 1988) confirms that under severe burning conditions, fires in untreated conifer stands commonly run through the crowns and are uncontrollable, with fireline intensities exceeding 1,000 BTUs per foot per second. After fuel reduction, the same burning conditions produce a surface fire that can be suppressed with hand tools (an intensity of less than 100 BTUs per foot per second). Coulter (1980) and Schmidt and Wakimoto (1988) give detailed recommendations for such fuel treatments. For example, shaded fuel breaks (open stands with minimal understory or surface fuels) can be created by treating fuels in a 1/8- to ¼-mile-wide forest belt around homes and developments. Breaks in fuel continuity lower the intensity of an approaching wildfire and aid suppression. Costs often can be partially offset with revenue generated from the sale of various forest products.

Fuels management forestry protects both natural resources and homes. Considering the high economic and social values at risk to wildfire in Zone III, intensive forestry to control fuels and maintain esthetic values is especially relevant to public needs. Small landowners need to be involved in fuels management. Educational efforts aimed at these owners should focus on

- the natural cycle of vegetation growth and fuel accumulation in forests and shrublands;
- the high values at risk in these natural "fire environments";
- how fuel accumulation invites catastrophic fires;
- how all the public pays for wildfire damage to homes through averaged insurance rates and disaster relief; and

- the effect of intensive forestry in reducing wildfire threats while enhancing forest health and esthetics.

The time is right for foresters and wildland fire specialists to demonstrate how innovative forest management can provide an ecologically sound and socially acceptable solution to threatened developments and natural resources in forested areas. In a larger sense, this is also a chance to demonstrate the adaptability of forestry and its value in promoting and perpetuating healthy forests. Foresters cannot afford to ignore this opportunity to serve society.

Literature Cited

Anderson, H. E., and J. K. Brown. 1988. Fuel characteristics and fire behavior considerations in the wildlands. P. 124-30 in Protecting people and homes from wildfire in the interior West proceedings of the symposium and workshop, W. C. Fischer and S. F. Arno, comps. USDA For. Serv. Gen. Tech. Rep. INT-251.

Arno, S. F., and R. H. Wakimoto. 1988.Fire ecology of vegetation common to wildland home sites. P. 118-23 in Protecting people and homes from wildfire in the interior West: proceedings of the symposium and workshop, W. C. Fischer and S. F. Arno, comps. USDA For. Serv. Gen Tech. Rep. INT-251.

Coulter, J. B. 1980. Wildfire safety guidelines for rural homeowners. Colo. State For. Serv., Fort Collins. 23 p.

Fisher, W. C., and S. F. Arno, comps. 1988. Protecting people and homes from wildfire in the interior West: proceedings of the symposium and workshop. USDA For. Serv. Gen. Tech Rep. INT-251. 213 p.

National Fire Protection Association. 1987 Wildfire strikes home! The report of the national wildland/urban fire protection conference. Natl Fire Prat. Assoc., Batterymarch Park, Quincy, MA. 90 p.

Nelson, T. C. 1979. Fire management policy in the national forests—a new era. J. For. 77:723-25.

Reeves, H. C. 1989. A better understanding of fire management. J. For. 87(2):inside back cover.

Schmidt, W. C., and R. H. Wakimoto. 1988. Cultural practices that can reduce fire hazards to homes in the interior West. P. 131-41 in Protecting people and homes from wildfire in the interior West: proceedings of the symposium and workshop, W. C. Fischer and S. F. Arno, comps. USDA For. Serv. Gen. Tech. Rep. INT- 251.

Wright, H. A., and A. W. Bailey. 1982. Fire ecology: United States and southern Canada. John Wiley & Sons, New York. 501 p.

Using Fire to Increase the Scale, Benefits, and Future Maintenance of Fuels Treatments

2012. *Journal of Forestry*. Vol 110. No. 7. 392–401.

Malcolm North, Brandon M. Collins, and Scott Stephens

The USDA Forest Service is implementing a new planning rule and starting to revise forest plans for many of the 155 National Forests. In forests that historically had frequent fire regimes, the scale of current fuels reduction treatments has often been too limited to affect fire severity and the Forest Service has predominantly focused on suppression. In addition to continued treatment of the wildland urban interface, increasing the scale of lowand moderate-severity fire would have substantial ecological and economics benefits if implemented soon. We suggest National Forests identify large contiguous areas to concentrate their fuels reduction efforts, and then turn treated firesheds over to prescribed and managed wildfire for future maintenance. A new round of forest planning provides an opportunity to identify and overcome some of the current cultural, regulatory, and institutional barriers to increased fire use that we discuss.

Keywords: fire policy, fire suppression, forest restoration, Forest Service planning rule, managed wildfire, Sierra Nevada

Affiliations: Malcolm North (mnorth@ucdavis.edu) is research scientist, USDA Forest Service, PSW, 1731 Research Park Drive, Davis, CA 95618. Brandon M.Collins (bmcollins@fs.fed.us), postdoctoral research, National Interagency Fire Center, USDA Forest Service, PSW, 1731 Research Park Drive, Davis, CA 95618. Scott Stephens (sstephens@berkeley.edu) is associate professor, Department of Environmental Science Policy and Management, 145 Mulford Hall, University of California, Berkeley, CA 94720.

Acknowledgments: The authors are grateful to Aaron Bilyeu and Jay Miller, Forest Service Region 5 Remote Sensing, for providing estimates of burn acreage and severity levels, Rob Griffith, Joe Sherlock, and Elizabeth Wright of Forest Service Region 5 Regional Office for providing acreage and cost data, Gus Smith of Yosemite N.P., Karen Fogler and Ben Jacobs of Sequoia/Kings Canyon N.P. for providing wildfire and prescribed fire acreage and cost estimates. The authors also thank Marc Meyer, Forest Service Southern Sierra Province Ecologist and three anonymous reviewers for their helpful reviews of an earlier version of this manuscript.

The USDA Forest Service (2012a) has adopted a new planning rule and is beginning to revise forest most of the National Forests, many of which are operating under 20–30 yr old plans. The planning rule directs responsible officials to "consider opportunities to restore fire adapted ecosystems and for landscape scale restoration" (Federal Register 77(68): 21174). Three of the eight forests nationally that will lead with plan revisions under this new rule are in California's Sierra Nevada (Sierra, Sequoia, and Inyo National Forests). In these forests, as in much of the western United States, fuels reduction has been a priority as the size and severity of wildfires has been increasing (Miller et al. 2009). However, both the scale and implementation rate for fuel treatment projects is well behind what is necessary to make a meaningful difference across landscapes (USDA Forest Service 2011). This issue is particularly relevant as wildfire size and intensity are projected to increase in many parts of the Sierra Nevada based on climate modeling (Lenihan et al. 2008, Westerling et al. 2011). It is almost an axiom in California forest management that given many existing restrictions, prescribed and managed wildfires will never be practical on a large scale for fuels reduction treatments (Quinn-Davidson and Varner 2011). California has some of the most restrictive air quality regulations in the country, a relatively high density of rural homes surrounded by flammable vegetation, extremely dry conditions during periods when prescribed fire could be used, and rugged topography that challenges containment efforts. Prescribed and managed wildfire do not generate revenue and therefore cannot cover management expenses, and managers may be liable if a fire escapes containment and damages property or inflicts injury. All of these concerns contribute to the general underuse of fire as a tool for both fuels reduction and forest restoration. We believe, however, much of the current approach to managing frequent-fire forests responds disproportionately to these immediate constraints and this approach is making it more difficult to reach long-term ecological and economic goals (Collins and Stephens 2007).

In this paper, we examine why managed wildfire and prescribed fire may be a more successful means of changing the scale and benefits of fuels treatment for fire-dependent forest ecosystems. We first estimate what the historic levels of burning may have been in California's Sierra Nevada under an active fire regime and compare this estimate to the scale of current fuels treatment efforts. We then discuss how ecological and economic benefits decline as fuels reduction is postponed. In an effort to overcome current constraints, we use "stretch goals" and "backcasting" (Manning et al. 2006) methods suggested by some restoration ecologists to identify a desired condition (i.e., greater fire use) and that work back to the means by which forest managers might get to this future goal. We conclude by examining some of the perceived constraints on increasing fire use and what changes may help diminish these constraints. It has been suggested that current fire policies are triage that treat the consequences of fire suppression without proactively focusing on redirecting fire to an ecologically beneficial role (Weatherspoon and Skinner 1996, Stephens and Ruth 2005). Increased prescribed burning and managed wildfire use may in part help effect some of this needed change.

In this discussion, we want to immediately make two distinctions about forest use areas and fire severity. What we focus on in this paper is the forest outside the wildland urban interface (WUI) (defined in the Federal Register as the area "where humans and their development meet or intermix with wildland fuel"). In the WUI, fire containment and suppression must be the primary goal of fuels treatments (Moghaddas and Craggs 2007). Outside the WUI, however, fuels treatments "should focus on creating conditions in which fire can occur without devastating consequences" (Reinhardt et al. 2008, p. 1998). Current policies often conflate fuels treatment with suppression or containment, when outside the WUI a more useful objective may be reducing adverse fire effects and intensity rather than occurrence and size (SNEP 1996, Stephens and Ruth 2005, Reinhardt et al. 2008). This raises the second distinction we want to emphasize. We focus on the benefits of restoring historic patterns of low and moderate-severity fire to forests that historically had frequent fire. We are not suggesting all fire is beneficial, particularly many modern wildfires that can have large areas burned at high severity. Outside of the WUI, management that emphasizes suppression ignores the inevitable consequence that reducing burn area in the present is counter productive in the long run. Inevitably these forests burn, and the longer that fire is excluded, the greater the likelihood that fire severity will increase and have large-scale adverse impacts (Biswell 1989, Marlon et al. 2012). The current priority and pace of fuels treatments outside the WUI is unlikely to significantly influence fire intensity and severity.

Current Fuels Treatments Compared to Historic Burn Acreage

Although recreating historic fire regimes may not be practical, understanding the extent of historic fire can give some general bounds on the level of fuels reduction that Sierra Nevada forests evolved with. How much Sierra Nevada forest would the Forest Service (FS) and National Park Service (NPS) need to treat each year to maintain the level of fuels reduction that forests experienced with an active (pre-1850) fire regime? We have included the NPS to compare with the FS because of its proportionally greater use of prescribed fire and managed wildfire to achieve land management objectives.

Using a GAP analysis that identified the acreage and agency ownership of different forest types in the Sierra Nevada (Davis and Stoms 1996), we calculated how much acreage might have historically burned each year by forest type. GAP analyses are used to identify how plant communities are distributed between different ownerships and aid in identifying where there are 'gaps' in conserving biodiversity (Scott et al. 1993). We loosely grouped the forest types into two categories; (1) active management for those forests more often outside of wilderness on Forest Service land and in the more accessible front country in the Sierra Nevada's

Management and Policy Implications

With less than 20% of the Sierra Nevada's forested landscape receiving needed fuels treatments, and the need to frequently re-treat many areas, the current pattern and scale of fuels reduction is unlikely to ever significantly advance restoration efforts. One means of changing current practices is to concentrate large-scale fuels reduction efforts and then move treated areas out of fire suppression into fire maintenance. A fundamental change in the scale and objectives of fuels treatments is needed to emphasize treating entire firesheds and restoring ecosystem processes. As fuel loads increase, rural home construction expands, and budgets decline, delays in implementation will only make it more difficult to expand the use of managed fire. Without proactively addressing some of these conditions, the status quo will relegate many ecologically important areas (including sensitive species habitats) to continued degradation from either no fire or wildfire burning at high severity.

Table 1. Forest type, total area, mean and high (mean of highest quartile) historic fire return interval (HFRI), fractional ownership, and approximate area that, on average, would have burned annually in the Sierra Nevada for the Forest Service (FS) and National Park Service (NPS) using the mean and high HFRI. The extent of the Sierra Nevada is the Jepson (Hickman 1993) definition, which is the area from the north fork of the Feather River south to Isabella Lake. Forest types are grouped into active and passive (forest types more often located in FS wilderness or NPS "back country") management.

| | | HFRI[a] | | Forest Service | | | | National Park Service | | | |
Forest type[b]	Area (ac)	Mean (yr)	High (yr)	Ownership	Area (ac)	Mean HFRI (ac/yr)	High HFRI (ac/yr)	Ownership	Area (ac)	Mean HFRI (ac/yr)	High HFRI (ac/yr)
Mix. conifer	1,466,539	12	25	0.62	909,254	75,771	36,370	0.05	73,327	6,111	2,933
West-side ponderosa	1,087,734	5	12	0.53	576,499	115,300	48,042	0.08	87,019	17,404	7,252
Lwr cismon. mix. con-oak	1,046,221	10	30	0.46	481,262	48,126	16,042	0.04	41,849	4,185	1,395
Jeff. pine-fir	730,428	8	25	0.8	584,342	73,043	23,374	0.09	65,738	8,217	2,630
Jeffrey pine	484,563	6	20	0.75	363,422	60,570	18,171	0.13	62,993	10,499	3,150
East-side ponderosa	398,819	5	15	0.76	303,103	60,621	20,207	0	0	0	0
Black oak	268,598	10	25	0.6	161,159	16,116	6,446	0.03	8,058	806	322
White fir	133,434	25	45	0.7	93,404	3,736	2,076	0.06	8,006	320	178
Aspen	24,463	30	90	0.89	21,772	726	242	0.02	489	16	5
Sequoia-mix con.	17,544	15	20	0.31	5,439	363	272	0.52	9,123	608	456
Active Man. Total	5,658,343				3,499,655	454,371	171,241		356,602	48,166	18,321
Red fir	838,905	45	90	0.61	511,732	11,372	5,686	0.3	251,671	5,593	2,796
Lodge. pine	532,748	30	110	0.6	319,649	10,655	2,906	0.42	223,754	7,458	2,034
Red fir-west. white p.	393,877	50	135	0.75	295,408	5,908	2,188	0.18	70,898	1,418	525
Whitebark p. mtn hemlock	93,404	85	180	0.62	57,910	681	322	0.37	34,559	407	192
Whitebark & lodge. pine	92,168	40	165	0.86	79,265	1,982	480	0.12	11,060	277	67
Up cismon. mix. con-oak	64,493	15	45	0.48	30,957	2,064	688	0.14	9,029	602	201
Foxtail pine	58,810	50	150	0.21	12,350	247	82	0.77	45,284	906	302
Whitebark p.	54,115	65	200	0.68	36,798	566	184	0.31	16,776	258	84
Passive Man. Total	2,128,519				1,344,068	33,475	12,536		663,031	16,918	6,201
All Man. Total	7,786,862				4,843,723	487,846	183,778		1,019,633	65,084	24,522

[a] Historic fire regime interval based on all studies cited in three sources with literature reviews, Stephens et al 2007, Safford and van de Water 2012, and the fire effects information database: www.fs.fed.us/database/feis/plants/tree/.
[b] Forest types based on Davis and Stoms 1996 for all forest types with > 10,000 ac.

two National Parks, Yosemite and Sequoia/ Kings Canyon, and (2) passive management for the other forest types (Table 1). We estimated the acreage that would annually burn in each forest type using two values of historic fire return intervals (HFRI). We calculated HFRI after reviewing two published studies (Stephens et al. 2007, van de Water and Safford 2011) that summarize information from hundreds of fire history studies, and the online fire effects information system (USDA Forest Service 2012b). We calculated the overall mean, and the mean of the highest quartile (hereafter referred to as high) of HFRI values (Table 1). We included the latter value as a very conservative estimate, one that managers might consider a minimal but approachable target given constraints on fuels reduction.

The analysis suggests the FS would need to reduce fuels annually on more than 487,000 ac/yr total (454,000 ac/yr in active management forest types) and more than 183,000 ac/yr total (171,000 ac/yr in active

management forest types) using the mean and high HFRI (Table 1), respectively, to approach historic levels. The NPS would need to reduce fuels annually on more than 65,000 ac/yr total (48,000 ac/yr in active management forest types) and more than 24,000 ac/yr total (18,000 ac/yr in active management forest types) using the mean and high HFRI, respectively (Table 1). Total acreage of FS lands is approximately five times that of the NPS in the Sierra Nevada, yet it has a much higher burn acreage total because it has ten times more acreage than the NPS in the forest types that generally have low HFRIs. In contrast, the FS has only twice the amount of acreage of the NPS in forest types generally at higher elevation with higher HFRIs.

Current annual fuels reduction on FS land averages 87,923 ac of which 28,598 ac is mechanical (33% of the total), 8,256 ac (9%) is prescribed fire and 51,069 ac (58%) is wildfire (Table 2). Combining both National Parks, the

Table 2. Average annual area and cost for mechanical treatment, prescribed fire, and wildfire control for Forest Service, Yosemite N.P. and Sequoia/Kings Canyon N.P lands. In the area columns, numbers in parentheses are years of record and in the cost columns are the minimums to maximums reported. All cost values have been standardized to 2012 dollars. Cost ranges should be treated as rough estimates as accounting practices vary within and between agencies.

| | Mechanical | | Rx burn | | Wildfire | | |
	Area (ac)	Cost ($/ac)	Area (ac)	Cost ($/ac)	Area (ac)	Cost ($/ac)	Tot. area (ac)
Forest Service[a]	28,598 (2004–2011)	$565 ($252–1077)	8,256 (2004–2011)	$145 ($72–619)	51,069 (1986–2010)	$830 ($746–28,834[b])	87,923
National Park	132 (2004–2011)	N/A	2,803 (1970–2011)	$206[3] ($153–458)	8,344 (1970–2011)	$496[c] ($413–2,063)	11,279

[a] For the Sequoia, Sierra, Stanislaus, Eldorado, Tahoe, Plumas, and Inyo National Forests and the Lake Tahoe Basin Management Unit.
[b] Detailed data for 2001–2010. Cost range is lowest for the largest fire size category (>2,000 ha) and highest for the smallest category (0–0.1 ha).
[c] Cost estimates only available from Sequoia/Kings Canyons N.P. For prescribed burning, the base cost is for using a N.P. crew, with costs increasing for using a contract crew. For wildfire, cost increases moving from accessible front country to backcountry wilderness.

average annual fuels reduction is 11,279 ac/yr of which 132 ac (1%) is mechanical, 2803 ac (25%) is prescribed fire, and 8,344 ac (74%) is wildfire (Table 2). Combining mechanical and prescribed fire treatments with wildfire, fuels are annually reduced on 18 or 48% (mean and high HFRI value, respectively) of the acreage that historically burned on FS land. Treatments and wildfire are reducing fuels on 17 or 46% (mean and high HFRI value, respectively) of NPS land. These estimates suggest both agencies have similar ratios of fuels reduced acreage to historic burned acreage, although by different means. For both agencies wildfire is the largest fuels reduction treatment, but often with much different effects. With the Forest Service's priority on suppression, which has been highly effective under moderate weather conditions (Finney et al. 2007), wildfires have escaped primarily under more extreme weather conditions. FS wildfires produce a greater proportion of high severity (33%) than on NPS land (15%, data from Yosemite NP only) (Miller et al. 2009, Thode et al. 2011, Miller, J.D. 2012, unpublished data), where fires more often burn under a wider range of weather conditions (Collins et al. 2007). Previous management practices (extensive timber harvesting, livestock grazing etc.), and a higher proportion of fuel-productive forest types may also contribute to these differences in proportions of high severity. On both FS and NPS lands, current fuels reduction acreage is substantially below historic levels, relegating most areas to accumulating high fuel loads. This increased fuel loading, however, does not necessarily produce higher fire severity if, following the NPS example, fires are allowed to burn under moderate weather conditions.

Ecological and Economic Effects of Current FS Fuels Reduction Practices

Ironically, current FS practices intended to protect resources identified as having high ecological value often put them at a greater risk of high-severity fire. A policy focused on suppression, which ultimately results in greater wildfire intensity, means that fuels reduction becomes the principle method of locally affecting fire behavior and reducing severity (Collins et al. 2010). Forest areas identified as having high conservation value, such as riparian conservation areas (van de Water and North 2010, 2011) and protected activity centers (PAC) for threatened and sensitive wildlife (North et al. 2010) often have management restrictions and higher litigation potential, resulting in minimal or no fuels reduction treatment. Stand conditions in these protected areas often consist of multilayered canopies with large amounts of surface fuel, resulting in increased crown fire potential (Spies et al. 2006, Collins et al. 2010). Following two particularly high-intensity 2007 wildfires in the Sierra Nevada (Angora and Moonlight), riparian and PAC areas had some of the greatest percentage of high-severity effects of any area within the fire perimeters (Dailey et al. 2008, Safford et al. 2009) (Figure 1). In contrast, low- and moderate-severity wildfire and prescribed burning in Yosemite N.P. maintained habitat characteristics and density of California spotted owls (*Strix occidentalis occidentalis*) in late successional montane forest (Roberts et al. 2011) (Figure 1).

This unintended consequence of suppression-focused fire policy is not limited to forest with special designations. A particular problem in many productive, frequent-fire forests is the high duff accumulations that develop around large old trees in the absence of fuels reduction (Figure 1). Long duration smoldering burns can kill even large trees

with thick bark (Ryan and Frandsen 1991, Hungerford et al. 1994). An analysis of factors associated with increased mortality found that when duff layers exceeded 5 in., the probability of mortality significantly increased, limiting conditions under which fire can burn without significant large-tree mortality (Hood 2010). The longer that fire is kept out of many productive forests, the greater the likelihood that burn intensity and duration will be higher than desired, impacting ecosystem services and potentially killing many of the large trees that are highly valued.

Of the three principle means of fuels reduction, mechanical, prescribed burning, and wildfire, the latter is often the most expensive in California. We surveyed Forest Service Region 5 and National Park Service personnel in Yosemite and Sequoia/Kings Canyon for cost estimate records (Table 2). [1] Costs should be treated as only general

Figure 1. Top left is a California spotted owl Protected Activity Center (PAC) following the 2007 Moonlight wildfire. Top right is a Jeffrey pine with a thick duff layer accumulation around its bole in fire-suppressed mixed conifer. Bottom is a California spotted owl nest in a mixed-conifer forest that burned in a prescribed fire with mixed fire severity in 1997 in Yosemite N.P. Note the nest (shown by arrow) is in an area that burned at low severity and has high canopy closure. The nest is adjacent (<50 ft away) to an area (left one third of the photo) with lower canopy closure that experienced moderate fire severity. (Photo credit: Stephanie Eyes).

estimates because accounting practices vary within and between agencies. The general trends, however, are instructive for identifying factors that increase costs and the rough differences between different treatments. Prescribed fire tends to be the least expensive per acre with costs decreasing as the size of treated area increases. Forest Service mechanical treatments costs vary widely but costs on average were 3.5 times higher than prescribed fire in large part due to expensive service contracts for removal of small, noncommercial biomass. Wildfire costs were highest but vary tremendously between burns. In general, costs per acre increased as access became more difficult but decreased with fire size.

These estimates are fairly consistent with published studies (Rideout and Omi 1995, González-Caban 1997, Cleaves et al. 1999, Butry et al. 2001, Berry and Hesseln 2004, Berry et al. 2006, Mercer et al. 2007). In one study examining Forest Service lands burned by wildfire and their suppression costs, total costs had high annual variability correlated with total area burned, while cost per acre from 1980 on were fairly consistent at about $760/ac (Calkin et al. 2005). In another study in southwest forests, the authors estimated that avoided future wildfire suppression costs justified spending $271 to $684/ac for fuels reduction (Snider et al. 2006). However, costs can vary substantially between regions. For Region 5, which includes the Sierra Nevada, wildland fire suppression costs averaged $2,567/ac for the period of 1995–2004 (Gebert et al. 2007). Factors having the largest influence on suppression cost were fire intensity (as measured by flame length), area burned, total housing value within 20 miles of the ignition and the percent of private land within the fire perimeter (Gebert et al. 2007, Liang et al. 2008). In addition, suppression costs in Region 5 may be associated with factors related to California's population size. A

recent paper suggested newspaper coverage and political pressure can substantially increase wildfire suppression costs (Donovan et al. 2010).

An additional economic consideration for the FS is that mechanical treatment costs are likely to increase. In many forests after decades of fire suppression fuels reduction costs, particularly for service contracts, can be offset with the removal of some trees large enough to have commercial value (Hartsough 2003). Second-entry treatments for maintenance of fuels reduction will probably have higher costs when commercial size trees are rare or no longer present. With current or even shrinking FS budgets, the scale of mechanical treatments may dramatically decrease for maintenance of areas previously treated.

In the future, maintenance of existing fuels reduction treatments could eventually subsume the entire treatment effort such that some proportion of the forest is never treated and always has uncharacteristically high fuel loads (hereafter, the backlog). We roughly calculated this backlog based on some assumptions given data limitations. We assumed mechanical and prescribed fire treatment would all occur in active management forest types and be proportional to each forest type's percentage of the Forest Service's total acreage. For wildfire we used the proportions of total wildfire acreage burned by forest type during 1984 –2004 (Miller et al. 2009; Table 1). Using criteria for Forest Condition Class 2 (Barrett et al. 2010), we assumed a forest would accumulate uncharacteristically high fuel loads if it were not treated within a period equal to twice the mean HFRI (Table 1) (Caprio et al. 2002). Using these proportions and current fuels reduction levels (Table 2), we calculated the total amount of acreage in each forest type that would have some form of fuels reduction (i.e., mechanical, prescribed burning, and wildfire) before treatments would have to "start over again." We then subtracted this amount from the forest type's total acreage to calculate backlog. We estimate that at current rates the deficit of forestland "in need" of treatment would be approximately 2.9 million ac (60%) of FS acreage in the Sierra Nevada, of which 1.7 million ac (60% of the backlog) are ponderosa and Jeffrey pine dominated forest types (Figure 2). We believe this is a very conservative estimate because it assumes that mechanical, prescribed fire, and wildfire areas never overlap.

An often-cited critique of fuels treatments is that the probability of wildfire burning a treated area within the expected lifespan of the treatment is so low that treatment costs and potential impacts on forest resources (e.g., carbon, watercourses, etc.) are rarely justified (Rhodes and Baker 2008,

Campbell et al. 2011). Indeed, in California fuels treatments designed for the suppression and containment of wildfire may rarely be economical except in the WUI or areas with known high ignition probabilities (i.e., road corridors). However, the cost-benefit ratio improves if the primary goal of fuels treatment is to reduce fire severity, not size, which is most effective when treatments are arranged strategically across a landscape (Finney et al. 2007). In general managed wildfire does fuel reduction "work" at a lower cost per acre than mechanically treating or suppressing the fire. In the long-term, however, the ecological and economic benefits of managed wildfire are only realized if the burned area is removed from the land base that requires suppression and in the future is maintained by fire.

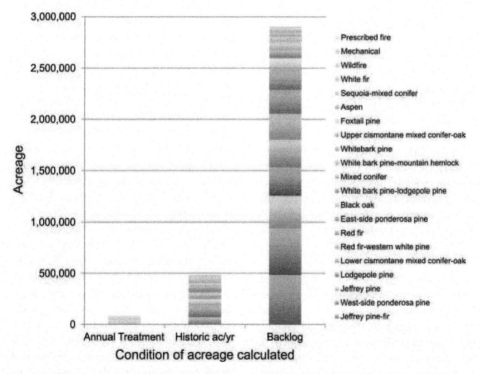

Figure 2. Histogram of the acreage of current annual fuels reduction by type, historical fuels reduction from wildfire by forest type and backlog by forest type. Backlog is a conservative estimate of the acreage that would always have uncharacteristically high fuel loads at current rates of fuels reduction and wildfire burning.

Stretch Goals and Backcasting

Collectively, current trends suggest the Forest Service needs a two- to fivefold increase in its annual fuels reduction acreage, is reducing the potential ecosystem services of untreated forest if burned by wildfire, and is engaged in a triage fuels reduction policy weighted toward suppression and high upfront, emergency costs. It is difficult to see how the pace, scale and economics of current practices can improve.

Frustrated by the ad hoc and small scale of many restoration projects, Manning et al. (2006) suggested using stretch goals and backcasting methods to significantly increase the scale and coordination of restoration efforts for whole ecosystems. A stretch goal is when an objective cannot be achieved by incremental improvements and requires a significant change in methods. Stretch goals are ambitious long-term targets used to generate innovation for achieving outcomes that currently seem impossible (Manning et al. 2006). Backcasting is where the stretch goal's desired condition is visualized and then a pathway to that condition is worked out retrospectively. The features of problems for which backcasting is well suited are when the problem is complex, there is a need for major change and marginal changes will be insufficient to solve a problem, dominant trends are part of the problem, and timescales are long enough to give considerable scope for deliberate choice (Dreborg 1996).

A hypothesized stretch goal for Sierra Nevada forest management is to restore as much ecologically beneficial fire into the landscape as possible (Miller et al. 2012). Using backcasting, the most practical means of getting to that desired endpoint is to significantly expand prescribed fire and managed wildfire. To meet this stretch goal we suggest that forest plans consider a new approach in fuels reduction policy: some treatments should be scaled and located with the intent of treating an entire fireshed and then converting that area to future management and maintenance through managed wildfire and prescribed fire. A fireshed has been defined as a contiguous area with similar fire history and problem fire characteristics where a coordinated suppression effort would be most effective (Ager et al. 2006, Bahro and Perrot 2006, Anonymous 2010). In California, efforts to identify firesheds have noted, ". . . boundaries are also influenced by the values they contain and by fire management opportunities" (Bahro et al. 2007). The opportunity in the firesheds we propose is to identify areas where fire is not suppressed but is restored as an active ecological process. Forest plans have the opportunity to identify these areas, establish criteria for how they will be treated, and under what conditions they will be allowed to burn. This is important because most forests will need more than one treatment before their burn window can be broadened. For example, after a long hiatus fire reintroduction can leave many small dead trees, that after toppling add to surface fuel loads (Skinner 2005). In these conditions fire hazard will remain elevated until decomposition or a second treatment reduces surface fuel loads.

This approach has several economic and ecological benefits. Many FS projects are 3,000 to 8,000 ac in size, whereas firesheds in California often encompass 50,000 to 100,000 ac or more (Bahro et al. 2007). This size would have economic efficiencies by providing an opportunity to bundle revenue-generating areas with lightly treated forests such as riparian and PAC areas that are revenue sinks (Hartsough 2003). For example, the FS is currently varying fuels treatments with topography to simulate the forest structure and pattern that might have historically been produced by an active fire regime (Skinner et al. 2006, North et al. 2009). Revenue from heavier thinning on upper slopes designed to restore low-density large pine conditions might be used to support hand thinning and/or prescribed burning that maintains higher canopy cover in the parallel track of forest in the drainage bottom (Figure 3). The larger scale of treatments and the practical need to spread them out over several years would make for a steady, more predictable flow of wood for local mills and potential biomass plants. Biomass use of small diameter fuels holds promise for improving the economics of fuels treatments. The lack of consistent biomass supply can limit development of processing infrastructure, however, large scale, long-term treatment planning can overcome some of these limitations (Hampton et al. 2011). Nor would this strategy just be a short-term boom, as it would take decades to widely implement fire-maintained firesheds, and once implemented there will still be a need for strategic mechanical fuels reduction. In the long run, creating burn-maintained firesheds could actually make headway in reducing the backlog of acreage needing treatment while realizing economic savings by substantially reducing future maintenance costs.

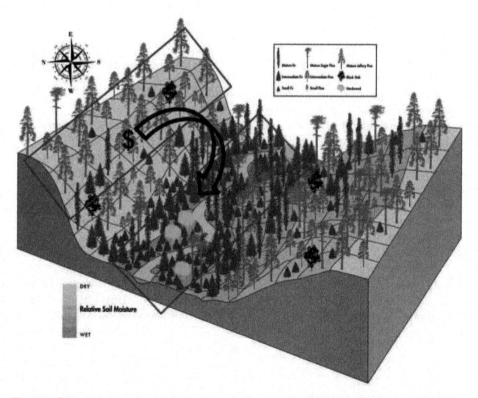

Figure 3. Example of how fuels treatments could be coupled to afford treating ecologically significant areas. The landscape schematic shows how variable forest conditions are produced by management treatments that differ by topographic factors such as slope, aspect, and slope position. Ridge tops and upper slopes have the lowest stem density and highest percentage of pine in contrast to riparian areas. Mid-slope forest density and composition varies with aspect: density and fir composition increase on more northern aspects and flatter slope angles. Revenue from heavier thinning on upper slope and ridge tops might be used to support minimal fuels treatment in riparian and PAC areas. (Figure drawn by Steve Oerding, U.C. Davis, Academic Technology Services).

Studies of different fuels treatments across a number of forest types historically associated with frequent fire consistently show the ecological benefits of fire and its essential role in ecosystem restoration (Stephens et al. 2009, 2012). Mechanical treatment alone may be able to mimic the live-tree structure of an active-fire forest condition (North et al. 2007), but it does not restore key ecological processes such as nutrient cycling (Wohlgemuth et al. 2006), understory (Wayman and North 2007, Webster and Halpern 2010) and microclimate (Ma et al. 2010) diversity, soil respiration patterns (Concilio et al. 2006), regeneration of fire-resistant tree species (Zald et al. 2008) or provision of habitat for some species (Hutto 1995, Saab and Powell 2005). Fire can restore some ecosystem benefits within a few applications. Forest structure (Taylor 2010) and understory conditions (Webster and Halpern 2010) may approach active fire regime conditions after two burns (or in some cases one moderate severity fire may be sufficient (Collins et al. 2011)). Recent research (Nesmith et al. 2011) also suggests that "despite restrictions" prescribed fire can produce similar patterns and effects on vegetation as low-intensity wildfire. Findings from these and other studies suggest that efforts to increase forest restoration and resilience need to incorporate fire. For this to occur at meaningful scales managed wildfire and prescribed fire needs to be a substantial component of the management portfolio.

Constraints on Fire Use

Escaped Fire Damage

Increased use of managed wildfire and prescribed fire is not without risk (e.g., 2012 Lower North Fork fire, Colorado; 2011 Margaret River fire, Western Australia; 2000 Cerro Grande fire, New Mexico). There are lessons, however, from these experiences that can help reduce the risk of escaped fire: (1) Prescribed fire, particularly for initial-entry burns, could be constrained to a relatively narrow set of fuel moisture and weather conditions. This would involve taking advantage of favorable conditions, whenever they occur (i.e., including nighttime and weekend opportunities). At present this can be difficult when land management agencies restrict work hours; (2) Minimize use of prescribed fire in areas surrounded by hazardous fuels. Prescribed fire units could be anchored by areas with low fire behavior potential (e.g., large rock outcrops, barren ridge tops, previous fuel treatments or wildfires). This involves developing a landscape strategy for fire use in the planning process. A new tool, the "Treatment Minimizer" in the software ArcFuels may help with this planning (Vaillant et al. in press). The stated justification for this tool is ". . . forest restoration should have the goal of creating the largest area within which fire behavior does not exceed thresholds that trigger suppression"; and (3) Burn large units (>1000 ac). There is very little to gain from burning small units relative to the risk. Revised forest plans can institute policy support for problems that will inevitably occur when fire use increases (Stephens et al. 2010).

Agency Culture

Although current policy recognizes the importance and need for managed wildfire (FWFMP 2001, USDA/USDI 2005, FWFMP 2009), studies have found very low rates of implementation. In 2004, land management agencies only let 2.7% of all lightning ignitions burn (NIFC 2006), consistent with a recent analysis in the Sierra Nevada that less than 2% of FS lands were burned under managed wildfire between 2001–2008 (Silvas-Bellanca 2011). The most significant factor associated with FS district rangers using managed wildfire was personal commitment, while the main disincentives were negative public perception, resource availability and perceived lack of agency support (Williamson 2007). For example, in Sierra Nevada National Forests, prescribed fire in mixed-conifer forests that produces >5% mortality of large overstory trees or patches of high severity has sometimes been considered as failing to meet objectives. Studies in an upper elevation mixed-conifer forest with a restored fire regime, however, suggest patches of considerable overstory mortality (>75%) do occur even under moderate weather and generally low fuel loads (Collins and Stephens 2010). These patches, however, are generally small (<10 ac) and collectively make up a relatively small portion of the landscape (<15%). The top changes suggested by managers to increase managed wildfire were increased training and education, institutional support, management flexibility and lands identified for managed wildfire (Doane et al. 2006). Implementation of the new planning rule and development of new forest plans provides an opportunity to address these problems and foster an agency culture that supports increased fire use.

Rural House Density
The presence and density of homes near a fire affect suppression costs (Liang et al. 2008), increase liability risk (White 1991, Czech 1996), and limit management options for managed wildfire use (Arno and Brown 1991). As more homes are built in the forest, the size and extent of the WUI increases. One analysis found the WUI in the United States expanded by more than 52% from 1970 to 2000 with the greatest increases occurring in the western United States (Theobald and Romme 2007). The human population in California's Sierra Nevada doubled between 1970 and 1990 and is projected to triple from 1990 to 2040 (Duane 1996). To date, however, much of that growth has been concentrated in areas with existing infrastructure (i.e., water

and power). Some projections of future development suggest new home construction will be more dispersed given technology improvements and increased interest in living 'off the grid' (White et al. 2009). Maps of current housing density based on Theobald's (2005) analysis of census data (Sierra Cascade Land Trust Council 2011) suggest unpopulated firesheds still exist in the Sierra Nevada, particularly in the northern and southern extent of the range. Fire-use area designation will become more difficult the longer it is postponed as more development occurs.

Air Quality
In California, air quality restrictions severely limit burn opportunities. In general, prescribed fire and managed wildfire ("natural ignitions that are managed for resource benefit") are subject to regulation by local air pollution control offices attempting to meet airshed standards under the Environmental Protection Agency's Title 17 (unmanaged wildfire is sometimes exempted from the standards on a case-by-case basis under the category of "exceptional events"). This has resulted in Yosemite and Sequoia/Kings Canyon N.P. being fined, or having permission for other planned projects denied when they did not suppress prescribed and managed wildfire even though they occurred in areas designated for fire use. On a per acre basis, however, emissions from an "escaped," unplanned, or high-severity wildfire can be substantially higher than occurs during managed wildfire or prescribed fire (Ahuju 2006). Fire-dependent forests will burn eventually, meaning the responsible choice is between periodic, lower concentrations of smoke in planned dispersal patterns or unplanned, heavy emissions where smoke drift and accumulation is uncontrolled. Current policy treats "unmanaged" wildfire occurrence and the resultant effects as 'an act of God' when human

management decisions and inaction have actually contributed to conditions that support large, severe fires. Changes in policy should be considered which acknowledge the inevitability of forest fire emissions and encourage responsible management actions that minimize harmful human exposure.

Conclusion

Region 5 of the Forest Service is embarking on developing their next round of forest plans designed to set the standards and guidelines for forestry practices for the next 10–20 years. Other Forest Service regions will likely be facing similar forest plan revisions in the near future and may look to Region 5 forests for guidance under the new programmatic planning rule. With less than 20% of the landscape that needs it receiving fuels treatments, and the need to re-treat many areas every 15–30 years depending on forest type, the current pattern and scale of fuels reduction is unlikely to ever significantly advance restoration efforts, particularly if agency budgets continue to decline. Treating and then moving areas out of fire suppression into fire maintenance is one means of changing current patterns. A fundamental change in the scale and objectives of fuels treatments is needed to emphasize treating entire firesheds and restoring ecosystem processes. As fuel loads increase, rural home construction expands, and budgets decline, delays in implementation will only make it more difficult to expand the use of managed fire. This approach may be criticized given current constraints but at least it could stimulate discussions between stakeholders, air quality regulators, and forest managers about current and future constraints on management options. Without proactively addressing some of these conditions, the status quo will relegate many ecologically important areas to continued degradation from fire exclusion. In some forests, revenue generated in the initial entry (Hartsough et al. 2008) may be the best opportunity to increase the scale and shift the focus of current fuels reduction toward favoring fire restoration.

Endnote

1. All dollar values reported in this paper have been standardized to 2012 dollars.

Literature Cited

Ager, A.A., M.A. Finney, and B. Bahro. 2006. Automating fireshed assessments and analyzing wildfire risk with ArcFuels. *For. Ecol. Manage.* 234S:215.

Ahuja, S. 2006. Fire and air resources. P. 444–465 in *Fire in California's Ecosystem*, Sugihara, N.G., J.W. van Wagtendonk, K.E. Shaffer, J. Fites-Kaufman, and A.E. Thode (eds.). University of California Press, Berkeley, CA.

California Board of Forestry and Fire Protection 2010. *The 2010 strategic fire plan for California. California Board of Forestry and Fire Protection: The Strategic Fire Plan.* Available online at www.bof.fire.ca.gov/board_committees/resource_protection_committee/current_projects/resources/2010_fire_plan_1-27-10 version.pdf; last accessed Feb. 15, 2012.

Arno, S.F., and J.K. Brown. 1991. Overcoming the paradox in managing wildland fire. *West. Wildl.* 17:40–46.

Bahro, B., and L. Perrot. 2006. Fireshed assessment. P. 454–456 in *Fire in California's Ecosystems*, Sugihara, N.G., J.W. van Wagtendonk, K.E. Shaffer, J. Fites-Kaufman, and A.E. Thode (eds.). University of California Press, Berkeley, CA.

Bahro, B., K.H. Barber, J.W. Sherlocks, and A. Yasuda. 2007. Stewardship and fireshed assessment: a process for designing a landscape fuel treatment strategy. Pages 41–54 in *Restoring fire-adapted ecosystems: Proceedings of the 2005 National Silviculture Workshop*, R.F. Powers (tech. ed.). USDA For. Ser., Rep. No. PSW-GTR-203.

Barrett, S., D. Havlina, J. Jones, W. Hann, C. Frame, D. Hamilton, K. Schon, T. Demeo, L. Hutter, and J. Menakis. 2010. *Interagency Fire Regime Condition Class Guidebook. Version 3.0*, Homepage of the Interagency Fire Regime Condition Class website, USDA Forest Service, US Department of the Interior, and The Nature Conservancy. Available online at www.frcc.gov; last accessed Aug. 3, 2012.

Berry, A.H., and H. Hesseln. 2004. The effect of the wildland-urban interface on prescribed burning costs in the Pacific Northwestern United States. *J. For.* 102:33–37.

Berry, A.H., G. Donovan, and H. Hesseln. 2006. Prescribed burning costs and the WUI: Economic effects in the Pacific Northwest. *West. J. Appl. For.* 21:72–78.

Biswell, H.H. 1989. *Prescribed Burning in California Wildlands Vegetation Management*. University of California Press, Berkeley, CA. 254 p.

Butry, D.T., D.E. Mercer., J.P. Prestemon, J.M. Pye, and T.P. Holmes. 2001. What is the price of catastrophic wildfire? *J. For.* 99:9–17.

Calkin, D.E., K.M. Gebert., J. Gre Jones, and R.P. Neilson. 2005. Forest Service large fire area burned and suppression expenditure trends, 1970–2002. *J. For.* 103:179–183.

Campbell, J.L., M.E. Harmon, and S.R. Mitchell. 2011. Can fuel-reduction treatments really increase forest carbon storage in the western US by reducing future fire emissions? *Front. Ecol. Env* 10:83–90.

Caprio, A.C., C. Conover, M. Keifer, and P. Lineback. 2002. Fire management and GIS: a framework for identifying and prioritizing fire planning needs. *Ass. Fire Ecol. Misc. Publ.* 1:102–113.

Cleaves, D.A., T.K. Haines, and J. Martinez. 1999. Prescribed burning costs: Trends and influences in the National Forest System. P. 277–288 in *Proc. of the Symposium on fire economics planning, and policy: Bottom lines*, A. González-Cabán and P.N. Omi (co-ords.). USDA For. Serv., Gen. Tech. Rep. No. PSWGTR-173.

Collins, B.M., and S.L. Stephens. 2007. Managing natural wildfire in Sierra Nevada wilderness areas. *Front. Ecol. Env* 5:523–527.

Collins, B.M., M. Kelly, J.W. Van Wagtendonk, and S.L. Stephens. 2007. Spatial patterns of large natural fires in Sierra Nevada wilderness area. *Lands. Ecol.* 22:545–557.

Collins, B.M., S.L. Stephens., J.J. Moghaddas, and J. Battles. 2010. Challenges and approaches in planning fuel treatments across fire-excluded forested landscapes. *J. For.* 108: 24–31.

Collins, B.M., and S.L. Stephens 2010. Stand-replacing patches within a mixed severity fire regime: Quantitative characterization using recent fires in a long-established natural fire area. *Lands. Ecol.* 25:927–939.

Collins, B.M., R.G. Everett, and S.L. Stephens. 2011. Impacts of fire exclusion and recent managed fire on forest structure in old-growth Sierra Nevada mixed-conifer forests. *Ecosphere* 2:51.

Concilio, A., S. Rhu, S. MA, M. North, and J. Chen. 2006. Soil respiration response to experimental disturbances over three years. *For. Ecol. Manage.* 228:82–90.

Czech, B. 1996. Challenges to establishing and implementing sound natural fire policy. *Renew. Res. J.* 14:14–19.

Dailey, S., J. Fites, A. Reiner, and S. Mori. 2008. *Fire behavior and effects in fuel treatments and protected habitat on the Moonlight fire.* Available online at www.fs.fed.us/adaptivemanagement/projects/FBAT/docs/Moonlight Final_8_6_08.pdf; last accessed Apr. 8, 2010.

Davis, F.W., and D.M. Stoms. 1996. Sierran vegetation: A gap analysis. P. 671–690 in *Sierra Nevada Ecosystem Project: Final Report to Congress.* Vol. II., University of California, Davis, CA, Centers for Water and Wildlands Resources, Wildland Resources Center Rep. No. 37.

Doane, D., J. O'Laughlin, P. Morgan, and C. Miller. 2006. Barriers to wildland fire use: A preliminary problem analysis. *Int. J. Wild.* 12: 36–38.

Donovan, G.H., J.P. Prestemon, and K. Gebert. 2010. The effect of newspaper coverage and political pressure on wildfire suppression costs. *Soc. Nat. Resour.* 24:785–798.

Dreborg, K.H. 1996. Essence of backcasting. *Futures.* 28:813–828.

Duane, T.P. 1996. Human settlement, 1850–2040. P. 235–360 in *The Sierra Nevada Ecosystem Project: Final Report to Congress.* Vol. II. Center for Water and Wildland Resources, U.C. Davis. Wildland Resources Center Rep. No. 37.

Federal Wildland Fire Management Policy (FWFMP). 2001. Review and update of the *1995 Federal Wildland Fire Management Policy,* Various governmental agency partners, Washington, DC. 78 p.

Federal Wildland Fire Management Policy (FWFMP). 2009. *Guidance for implementation of federal wildland fire management policy.* Various governmental agency partners, Washington DC. 20 p.

Finney, M.A., R.C. Seli, C.W. Mchugh, A.A. Ager, B. Bahro, and J.K. Agee. 2007. Simulation of long-term landscape-level fuel treatment effects on large wildfires. *Int. J. Wildland Fire.* 16:712–727.

Gebert, K.M., D.E. Calkin, and J. Yoder. 2007. Estimating suppression expenditures for individual large wildland fires. *West J. Appl. For.* 22:188–196.

González-Caban, A. 1997. Managerial and institutional factors affect prescribed burning costs. *For. Sci.* 43:535–543.

Hampton, H.M., S.E. Sesnie, J.D. Bailey, and G.B. Snider. 2011. Estimating regional wood supply based on stakeholder consensus for forest restoration in northern Arizona. *J. For.* 109: 15–26.

Hartsough, B. 2003. Economics of harvesting to maintain high structural diversity and resulting damage to residual trees. *West J. Appl. For.* 18:133–142.

Hartsough, B.R., S. Abrams, R.J. Barbour, E.S. Drews, J.D. Mciver, J.J. Moghaddas, D.W. Schwilk, and S.L. Stephens. 2008. The economics of alternative fuel reduction treatments in western United States dry forests: financial and policy implications from the National Fire and Fire Surrogate Study. *For. Econ. Pol.* 10:344–354.

Hickman, J.C. 1993. *The Jepson Manual of Higher Plants of California.* University of California Press, Berkeley and Los Angeles, CA. 1400 p.

Hood, S.M. 2010. *Mitigating old tree mortality in long-unburned, fire-dependent forests: a synthesis.* USDA For. Serv. Gen. Tech. Rep. No. RMRS-GTR-238. 71 p.

Hungerford, R.D., K.C. Ryan, and J.J. Reardon. 1994. Duff consumption: New insights from laboratory burning. P. 472–476 in *Proc. of the 12th International conference on fire and forest meteorology*; October 26–28, 1993, Jekyll Island, GA, Society of American Foresters, Bethesda, MD.

Hutto, R.L. 1995. The composition of bird communities following stand-replacement fires in northern Rocky Mountains (U.S.A.) conifer forests. *Conserv. Biol.* 9:1041–1058.

Lenihan, J.M., D. Backelet, R.P. Neilson, and R. Drapek. 2008. Response of vegetation distribution, ecosystem productivity, and fire to climate change scenarios for California. *Clim. Change* 87:S215–S230.

Liang, J., D.E. Calkin, K.M. Gebert, T.J. Venn, and R.P. Silverstein. 2008. Factors influencing large wildland fire suppression expenditures. *Int. J. Wildl. Fire* 17:650–659.

Ma, S., A. Concilio, B. Oakley, M. North, and J. Chen. 2010. Spatial variability in microclimate in a mixed-conifer forest before and after thinning and burning treatments. *For. Ecol. Manage.* 259:904–915.

Manning, A.D., D.B. Lindenmayer, and J. Fisher. 2006. Stretch goals and backcasting: Approaches for overcoming barriers to large scale ecological restoration. *Restor. Ecol.* 14: 487–492.

Marlon, J.R., P.J. Bartlein, D.G. Gavin, C.J. Long, R.S. Anderson, C.E. Briles, K.J. Brown, D. Colombaroli, D.J. Hallett, M.J. Power, E.A. Scharf, and M.K. Walsh. 2012. Long-term perspective on wildfires in the western USA. *Proc. Natl. Acad. Sci. U.S.A.* in press. doi: 10.1703/pnas.1112839109.

Mercer, D.E., J.P. Prestemon, D.T. Butry, and J.M. Pye. 2007. Evaluating alternative prescribed burning policies to reduce net economic damages from wildfire. *Am. J. Agric. Econ.* 89:63–77.

Miller, J.D., H.D. Safford, M. Crimmins, and A.E. Thode. 2009. Quantitative evidence for increasing forest fire severity in the Sierra Nevada and southern Cascade Mountains, California and Nevada, USA. *Ecosystems.* 12: 16–32.

Miller, J.D., C.N. Skinner, H.D. Safford, E.E. Knapp, and C.M. Ramirez. 2012. Trends and causes of severity, size, and number of fires in northwestern California, USA. *Ecol. Appl.* 22:184–203.

Moghaddas, J.J., and L. Craggs. 2007. A fuel treatment reduces fire severity and increases suppression efficiency in a mixed conifer forest. *Int. J. Wildl. Fire.* 16:673–678.

National Interagency Fire Center (NIFC). 2006. *Wildland fire statistics.* National Interagency Fire Center: Boise, ID. Available online at www.nifc.gov/fire_info/fire_stats.htm; last accessed Feb. 15, 2012.

Nesmith, J.C.B., A.C. Caprio, A.H. Pfaff, T.W. Mcginnis, and J.E. Keeley. 2011. A comparison of effects from prescribed fires and wildfires managed for resource objectives in Sequoia and Kings Canyon National Parks. *For. Ecol. Manage.* 261:1275–1282.

North, M., J. Innes, and H. Zald. 2007. Comparison of thinning and prescribed fire restoration treatments to Sierran mixed-conifer historic conditions. *Can. J. For. Res.* 37:331–342.

North, M., P. Stine, K. O'Hara, W. Zielinski, and S. Stephens. 2009. *An ecosystem management strategy for Sierran mixed-conifer forests.* USDA For. Serv., PSW Gen. Tech. Rep. No. PSW-GTR-220, 49 p.

North, M., P. Stine, W. Zielinski, K. O'Hara, and S. Stephens. 2010. Harnessing fire for wildlife. *Wildl. Prof.* 4:30–33.

Quinn-Davidson, L.N., and J.M. Varner. 2011. Impediments to prescribed fire across agency, landscape and manager: an example from northern California. *Int. J. Wildl. Fire* 21:210–218.

Reinhardt, E.D., R. Keane, D.E. Calkin, and J.D. Cohen. 2008. Objective and considerations for wildland fuel treatment in forested ecosystems of the interior western United States. *For. Ecol. Manage.* 256:1997–2006.

Rhodes, J.J., and W.L. Baker. 2008. Fire probability, fuel treatment effectiveness and ecological tradeoffs in western U.S. public forests. *Open For. Sci. J.* 1:1–7.

Rideout, D.B., and P.N. Omi. 1995. Estimating the cost of fuels treatment. *For. Sci.* 41:664–674.

Roberts, S.L., J.W. Van Wagtendonk, A.K. Miles, and D.A. Kelt. 2011. Effects of fire on spotted owl site occupancy in a late-successional forest. *Biol. Conserv.* 144:610–619.

Ryan, K.C., and W.H. Frandsen. 1991. Basal injury from smoldering fires in mature Pinus ponderosa Laws. *Int. J. Wildl. Fire.* 1:107–118.

Saab, V.A., and H.D.W. Powell. 2005. Fire and avian ecology in North America: Process influencing pattern. *Stud. Avian Biol.* 30:1–13.

Safford, H.D., D.A. Schmidt, and C.H. Carlson. 2009. Effects of fuel treatments on fire severity in an area of wildland-urban interface, Angora Fire, Lake Tahoe Basin, California. *For. Ecol. Manage.* 258:773–787.

Scott, J.M., F. Davis, B. Csuti, R. Hoss, B. Butterfield, C. Groves, H. Anderson, S. Caicco, F. D'Erchia, T.C. Edwards JR., J. Ulliman, and R.G. Wright. 1993. Gap analysis: a geographic approach to protection of biological diversity. *Wildl. Mono.* 123:1–41.

Sierra Cascade Land Trust Council. 2011. *Foothills Area Conservation Report.* Available online at www.sierracascadelandtrustcouncil.org/foothills-area-conservation-report/; last accessed Aug. 3, 2012.

Sierra Nevada Ecosystem Project (SNEP). 1996. *Sierra Nevada Ecosystem Project, Final Report to Congress*, Vol. I–IV. Centers for Water and Wildland Resources, University of California, Davis, CA.

Silvas-Bellanca, K. 2011. *Ecological burning in the Sierra Nevada: Actions to achieve restoration.* White paper of The Sierra Forest Legacy. Available online at www.sierraforestlegacy.org/; last accessed Feb. 24, 2012.

Skinner, C.N. 2005. Reintroducing fire into the Blacks Mountain Research Natural Area: effects on fire hazard. P. 245–257 in *Proc. of the Symposium on ponderosa pine: Issues, trends, and management.* Ritchie, M.W., D.A. Maguire, and A. Youngblood (eds.). USDA For. Serv. Gen. Tech. Rep. No. PSW-GTR-198.

Skinner, C.N., A.H. Taylor, and J.K. Agee. 2006. Klamath Mountains bioregion. P. 170–194 in *Fire in California ecosystems*, Sugihara N.S., J.W. van Wagtendonk, J. Fites-Kaufmann, K. Shaffer, and A. Thode (eds.). University of California Press, Berkeley, CA.

Snider, G., P.J. Daugherty, and D. Wood. 2006. The irrationality of continued fire suppression: An avoided cost analysis of fire hazard reduction treatments versus no treatment. *J. For.* 104:431–437.

Spies, T.A., M.A. Hemstrom, A. Youngblood, and S. Hummel. 2006. Conserving oldgrowth forest diversity in disturbance-prone landscapes. *Conserv. Biol.* 20:351–362.

Stephens, S.L., and L.W. Ruth. 2005. Federal forest-fire policy in the United States. *Ecol. Appl.* 15:532–542.

Stephens, S.L., R.E. Martin, and N.E. Clinton. 2007. Prehistoric fire area and emissions from California's forests, woodlands, shrublands, and grasslands. *For. Ecol. Manage.* 251: 205–216.

Stephens, S.L., J.J. Moghaddas, C. Edminster, C.E. Fiedler, S. Haase, M. Harrington, J.E. Keeley, E.E. Knapp,

J.D. Mciver, K. Metlen, C.N. Skinner, and A. Youngblood. 2009. Fire treatment effects on vegetation structure, and potential fire severity in western U.S. forests. *Ecol. Appl.* 19:305–320.

Stephens. S.L., C.I. Millar, and B.M. Collins. 2010. Operational approaches to managing forests of the future in Mediterranean regions within a context of changing climates. *Envi-ron. Res. Lett.* 5:024003.

Stephens, S.L., J.D. McIver, R.E.J. Boerner, C.J. Fettig, J.B. Fontaine, B.R. Hartsough, P. Kennedy, and D.W. Schwilk. 2012. Effects of forest fuel reduction treatments in the United States. *BioScience* 62:549–560.

Taylor, A.H. 2010. Fire disturbance and forest structure in an old-growth Pinus ponderosa forest, southern Cascades, USA. *J. Veg. Sci.* 21: 561–570.

Thode, A.E., J.W. Van Wagtendonk, J.D. Miller, and J.F. Quinn. 2011. Quantifying the fire regime distributions for severity in Yosemite National Park, California, USA. *Int. J. Wildl. Fire.* 20:223–239.

Theobald, D.M. 2005. Landscape patterns of exurban growth in the USA from 1980 to 2020. *Ecol. Soc.* 10:32.

Thoebald, D.M., and W.H. Romme. 2007. Expansion of the US wildland-urban interface. *Landsc. Urban Plann.* 83:340–354.

USDA Forest Service. 2011. *Region Five Eco-logical Restoration: Leadership Intent.* USDA Forest Service, Pacific Southwest Region. 4 p.

USDA Forest Service. 2012a. *The Forest Planning Rule.* Available online at www.fs.usda.gov/planningrule; last accessed Aug. 3, 2012.

USDA Forest Service. 2012b. *Tree List.* Available online at www.fs.fed.us/database/feis/plants/tree/; last accessed Aug. 3, 2012.

USDA/USDI. 2005. *Wildland Fire Use: Implementation Procedures Reference Guide.* National Wildfire Coordinating Group. National Interagency Fire Center: Boise, ID. 71 p.

Vaillant, N.M., A.A. Ager, J. Anderson, and L. Miller. 2012. *ArcFuels User Guide: for use with ArcGIS 9.X.*, USDA For. Serv. Gen. Tech. Rep. No. PNW-GTR. 225 p.

Van de Water, K., and M. North. 2010. Fire history of coniferous riparian forests in the Sierra Nevada. *For. Ecol. Manage.* 260:384–395.

Van de Water, K., and M. North. 2011. Stand structure, fuel loads, and fire behavior in riparian and upland forests, Sierra Nevada Mountains, USA; a comparison of current and reconstructed conditions. *For. Ecol. Manage.* 262:215–228.

Van de Water, K.M., and H.D. Safford. 2011. A summary of fire frequency estimates for California vegetation before Euro-American settlement. *Fire Ecol.* 7:26–58.

Wayman, R., and M. North. 2007. Initial response of a mixed-conifer understory plant community to burning and thinning restoration treatments. *For. Ecol. Manage.* 239:32–44.

Weatherspoon, C.P., and C.N. Skinner. 1996. Landscape-level strategies for forest fuel management. P. 1471–1492 in *Sierra Nevada Ecosystem Project: Final Report to Congress*, Vol. II. Centers for Water and Wildland Resources, University of California, Davis, Wildland Resources Center Report No. 37.

Webster, K.M., and C.B. Halpern. 2010. Long-term vegetation responses to reintroduction and repeated use of fire in mixed-conifer forests of the Sierra Nevada. *Ecosphere* 1:27.

Westerling, A.L., B.P. Bryant, H.K. Preisler, T.P. Holmes, H.G. Hidalgo, T. Das, and S.R. Shrestha. 2011. Climate change and growth scenarios for California wildfire. *Clim. Change* 109: S445–S463.

White, D.H. 1991. Legal implications associated with use and control of fire as a management practice. P. 375–384 in *High intensity fire in wildlands: Management challenges and options*, S.M. Hermann (ed.). Tall Timber Fire Ecology Conference Proc. 17, May 18–211989, Tallahassee, FL.

White, E.M., A.T. Morzillo, and R.J. Alig. 2009. Past and projected rural land conversion in the US at state, regional, and national levels. *Landsc. Urban Plann.* 89:37–48.

Williamson, M.A. 2007. Factors in United States Forest Service district rangers' decision to manage a fire for resource benefit. *Int. J. Wildl. Fire.* 16:755–762.

Wohlgemuth, P.M., K. Hubbert, and M.J. Arbaugh. 2006. Fire and physical environment Interactions: soil, water and air. P. 75–93 in *Fire in California's ecosystem*. Sugihara, N. G., J.W. van Wagtendonk, K.E. Shaffer, J. Fites-Kaufman, and A.E. Thode (eds.). University of California Press, Berkley, CA.

Zald, H., A. Gray, M. North, and R. Kern. 2008. Initial regeneration responses to fire and thinning treatments in a Sierra Nevada mixed- conifer forest, USA. *For. Ecol. Manage.* 256:168–179.

Part 4

Looking to the Future: Fire Education, Training, and Research Needs

Alistair M.S. Smith

This part of the anthology takes a retrospective look at the drivers behind advances in fire education, training, and research. This is accompanied by a forward-thinking glance at what may drive future advances over the next century. In the United States, wildland fire science has a strong history of innovation and these advances have, in large part, been due to a series of individuals willing to see past the cultural norm and conduct out-of-the-box investigations or simply to stand up and make powerful statements of why things should change. Clearly, this anthology cannot include all the innovators from the past century of wildland fire science education, training, and research; it, however, does capture some of the most influential voices that have set the stage for the next century of innovators to rise and to help solve the problems associated with wildland fire science.

One of the earliest and most influential innovators was Harry Gisborne, so it is fitting it is his paper on "Mileposts of Progress in Fire Control and Fire Research" that starts off this part of the anthology.

By way of introduction, Harry Gisborne's mark on wildland fire science and training in the United States was at its peak in the 1930s, but his influence on how we assess and approach fires persists to the present day. Although Gisborne would, in 1937, become the Chief of the Division of Forest Fire Research, he is widely remembered for the development in 1935 of a fire-danger meter, while at the Priest River Experimental Forest. The fire-danger meter was a visual device, about the same size of a pair of binoculars, which used weather and fuel data to assess fire risk. Gisborne's paper in the *Journal of Forestry* in 1942 was a reflective look at what he saw as the main milestones in wildland fire control during the preceding 35 years. At the time, the Editor in Chief of the *Journal of Forestry* hoped that his article would encourage "old-timers" in the discipline to provide suggestions on other significant events and innovations. Roy Headley rose to the challenge and provided a two-page afterword that clarified some points and provided additional details.

Gisborne opened his paper with a powerful statement, "I am convinced that failure in recent years to look at the whole problem first, and its parts second, had caused a large part of what we now call the fire problem". Regrettably, it can easily be argued that this fire problem still persists to the present day as a holistic system view of fires, where fuels, fires, other disturbances, climate, and humans are all intertwined, struggles to gain traction in wildland fire education, training, and research. After that opener, Gisborne presented an anthology of significant procedure and policy events each year from 1905, wherein he described the intended and unintended consequences that resulted. Gisborne's paper was very detailed and informative with many highlighted resources to pursue in more depth. His narrative was filled with passion and contains numerous powerful and insightful statements, clearly

emboldened through his considerable depth of personal and professional experience in wildland fire.

Early in his paper Gisborne included an acknowledgement of the former Chief Forester's, H.S. Graves, statements on the fire problem in the *1910 Forest Service Bulletin No. 82: Protection of Forests by Fire*. He also highlighted the application of radio by R.B. Adams on an operational fire in 1921 and referenced "A National Program of Forest Research" by E.H. Clapp that Gisborne remarked at the time was "so well appreciated that it needs no explanation." A notable highlight Gisborne describes in detail was the initiation of the Civilian Conservation Corps (CCC) in 1933, in which he noted that, in his opinion, the surge of methods and personnel brought about by the CCC necessitated the improved training of wildland firefighters. For those not familiar with the long-term impact of the CCC on present-day wildland fire management; many of the fire lookout towers and service buildings still in use today were built or improved during the CCC era that lasted until 1942.

Gisborne's narrative is not without wit and humor. His dismay at the inception of the 1935 10 a.m. policy was unmistakable in his writings, where he described it as "either a milepost or a tombstone on our 35-year road of progress." He had hoped it was only a temporary measure, but clearly feared the longevity of the policy apparent in his remarks at the time, "If, however, it has already become or ever does become the death knell of all previous objectives based on damage, then it rates a tombstone executed in the blackest of black granite."

These words now lead us to our second highlighted innovator, Harold Weaver and his paper titled, "Fire as an Ecological and Silvicultural Factor in the Ponderosa-Pine Region of the Pacific Slope" that was published in the *Journal of Forestry* in 1943.

For those unfamiliar with the work of Harold Weaver; he is widely recognized as one of the most influential pioneers in the use and application of prescribed burning. Weaver and his colleague and counterpart Harold Biswell, with whom he worked at the University of California, are widely considered the forerunners of prescribed burning in the western United States. Harold Weaver worked as a forester for the Bureau of Indian Affairs and conducted a series of prescribed burns on the Confederated Tribes of the Colville Reservation lands in Washington State. Many forestry and wildland fire science researchers are of the opinion that Weaver enjoyed a fair degree of independence through working at the Bureau of Indian Affairs that enabled him to proposed outlandish and innovative ideas around the use of prescribed fires that were in stark contrast to the staunch position of fire suppression and control employed by the United States Forest Service. Indeed, at the time Weaver's paper was considered so controversial that it was formally rebutted by the then Director of the United States Forest Service Fire Control Research unit, A.A. Brown; an individual who clearly did not share the progressive views of Gisborne. Ironically, the rebuttal ended with a challenge to Weaver to prove his claims that fire could be used as a silvicultural tool beyond removal of slash through more research; thankfully, the wider fire science and forestry research community took on this task and has ultimately proven Weaver's case hundreds of times over, upholding his once radical ideas beyond any reasonable doubt.

Weaver's paper provided evidence that the total suppression of forest fires in the ponderosa pine forests of the Rockies and the Cascades produced "undesirable" ecological effects. Weaver opened his paper with his observations of past fires on the landscape and his hypothesis that, "periodic fires, in combination frequently with pine-beetle attacks, and occasionally with other agencies, formerly operated to control the density, age classes, and composition of the ponderosa-pine stands." To many of the advocates of total fire suppression, such controversial comments were akin to heresy. However, Weaver did an unexpected thing; rather than just present his opinions, he synthesized available evidence.

Weaver starts his narrative with an explanation of the formation of fire scars and a synthesis of the tree-ring record studies that had been conducted in the ponderosa pine region, where he clearly demonstrated the frequency of past fires affecting the region. Weaver then presented evidence of how fire has impacted ponderosa pine stands and provided a discussion of the importance of fire on the ecology of these systems. Weaver noted that many foresters knew of fire's past role in these pine systems, but that they had a mistaken perception that fire kept those forests understocked and "worn down by the attrition of repeated light fires." Weaver closed his paper with a challenge to the "best minds in forest research" to solve the undesirable ecological impacts of fire exclusion on ponderosa pine systems. One of the most compelling and enduring aspects of Weaver's paper was the inclusion of photographs that created a visual record of the impact that total fire exclusion had on the ponderosa pine systems.

The third paper in this part takes us nearly 40 years into the future. This time it reflects a partnership by Stephen Barrett and Stephen Arno in their paper entitled, "Indian Fires as an Ecological Influence in the Northern Rockies" that was published in the *Journal of Forestry* in 1982.

Stephen Barrett was a forester from Kalispell, Montana, who widely wrote about the aboriginal use of fires within the United States. Stephen Arno was a former research forester with the United States Department of Agriculture Rocky Mountain Research Station. Together, their paper outlined the role of fire on Indian lands in the northwestern United States. In contrast to the earlier paper by Weaver, the very first sentence of the paper by Barrett and Arno highlights how perspectives regarding fire in this region had markedly changed, namely, "The importance of fire as an ecological disturbance in the Northern Rockies is well accepted." Into this context, Barrett and Arno provide a succinct but compelling synthesis on the widespread use and role of fires by Native Americans on the northern Rocky Mountain forests. The paper was a summary of a wider study that was presented in Barrett's MS thesis at the University of Montana. As with the earlier synthesis of Weaver, Barrett and Arno presented fire frequency data. However, they used this data to compare the frequency of fires prior to European settlement (pre-1860) compared to those fires produced from lightning ignitions during the fire exclusion period from 1931-1980. Another stark contrast that illustrates how times had changed from the comments of Weaver, Barrett and Arno concluded that "managers of these forests might consider prescribed underburning ... Such underburns, using appropriate techniques and prescriptions ... could reduce fuel and pathogen hazards, with little harm to pole and lathe crop trees of ponderosa pine, western larch, and Douglas-fir." The telling phrase is "using appropriate techniques and prescriptions" as it shows that the once controversial prescribed fires under Weaver had, by this time, become the accepted practice.

The penultimate paper in this part takes us to 10 years after the 1988 Yellowstone Fires, with "The Politics of Wildfire: Lessons from Yellowstone" by Pamela Lichtman that was published in the *Journal of Forestry* in 1998.

At the time of this article, Pamela Lichtman was a research associate at the Northern Rockies Conservation Cooperative in Jackson, Wyoming. This article was not only timely, given its publication 10 years after the 1988 fires in the Greater Yellowstone Ecosystem, but also of tremendous value because it covered topics that were generally ignored in the debate of what do about wildfires; namely, politics and public perceptions. The impact of Lichtman's paper was profound, as it effectively established a new subdiscipline of wildland fire science. Her ideas have now been adopted by other social scientists that have sought to understand the factors associated with public perception of wildland fires. Widely recognized names in wildfire public perceptions research like Sarah McCaffrey, whose early work included a *Journal of Forestry* article in 2004 entitled "Fighting Fire with Education" arguably entered the discipline following Lichtman's work. More recently, the highly innovative and groundbreaking scientist, Travis Paveglio, has picked up the torch to advance these research questions. For those yet unfamiliar with Paveglio's work, he recently was the author of "Categorizing the Social Context of the Wildland Urban Interface: Adaptive Capacity for Wildfire and Community "Archetypes" that was published in *Forest Science* in 2015 and "Understanding the Effect of Large Wildfires on Residents' Well-Being: What Factors Influence Wildfire Impact?" published in *Forest Science* in 2016.

Lichtman's paper begins with the two simple but hard-hitting questions: "Did fire policy fail in 1988? Did it irreparably damage our natural heritage and destroy commodities?" Lichtman noted that although the ecological impacts of the fires were inevitable, the political firestorm that followed the 1988 Yellowstone Fires could have been avoided. Lichtman starts her narrative with a call for an increased understanding of the public perceptions surrounding natural fires as a means to ensure public support of ecosystem-based land management policies. Lichtman presented a synthesis of the anti-fire sentiment that followed the 1988 Yellowstone Fires, including many dire predictions about the likely impact on tourism and wildlife; predictions that ultimately did not come to pass. Although the increase in national attention actually led to a peak in tourism following the fires, Lichtman notes that the negative public perceptions surrounding wildfires persist and seem to re-emerge with each new large wildland fire season. Near the end of her narrative, Lichtman made this powerful statement, "Understanding how this process occurs and managing it will require not only effective communication ... but also mutual understanding and dialogue with people who are directly affected by fire ..." and a suggestion on how to achieve it: "Well-designed participatory structures incorporating the best available techniques ... could increase effective participation by a region's residents." These suggestions have arguably acted as the lighthouse that has guided wildland fire science perceptions research that followed.

The final paper in this part entitled, "Challenges to Educating the Next Generation of Wildland Fire Professionals in the United States," is unique as it was authored by the Education Subcommittee of the Association for Fire Ecology. At the timing of this anthology, its lead author, Leda Kobzair, was the current President of the Association of Fire Ecology. This article was published in the *Journal of Forestry* in 2009. This paper exhibits a very different tone from the other contributions in this part, as it was predominately written by university academics who are covering the topic of fire education and training that has, and arguably still, rests in the hands

of the federal and state fire organizations. Consequently, the authors had to walk a very careful tightrope, as they clearly wanted to stand up and highlight the challenges with the current approaches, while presenting potential solutions and a path forward.

The narrative starts a little awkwardly with the authors arguably making an unforced error by seeking to assert their expertise over the established professionals by declaring that they were drawing on over 65 years of combined experience in education in wildland fire science. However, apart from Professors Morgan and Sugihara who worked for the Forest Service, all the other authors in 2009 were new to their respective faculty positions. The result was an initial 'we know best' attitude, something that never sits well with professionals who have decades of hands-on operational experience. To their credit, the authors followed this statement with an explanation that they did reach out to fire management professionals, conducted panels, and solicited formal feedback as part of their synthesis. After this rocky start, the authors recovered their poise and the narrative was able to make several strong and impactful statements. In the context of describing the broad challenges associated with an aging and retiring wildland fire science workforce, the authors asked the compelling question of, "How would we prepare the next generation of fire professionals if we were to start from scratch?" They contended that this question required them to "explore innovative solutions to the current challenges."

They argued that education and training in wildland fire management was shifting from solely fire suppression tactics to a system in which fire use and suppression were both options in a broader toolkit. The authors highlighted that there is a need for a new generation of fire professionals and introduces a new fire triangle to wildland fire science, namely Training, Experience, and Education. They also emphasized the need for education in tools and technology including models, geospatial datasets, and ecosystem analysis. The authors presented many of the challenges that inhibit a re-envisioning of the wildland fire education and training system in the United States. These included a lack of formal coordination between land management agencies and educational institutions, the overlapping of fire seasons and academic semesters and, most notably, that the National Wildfire Coordination Group (NWCG) does not recognize university courses unless they meet their specific curriculum and are taught by a NWCG qualified instructor. As noted by the authors, this latter point is a major roadblock as "Few available instructors meet both university and NWCG requirements." Also, if a university updates or expands on the NWCG course material, the courses are no longer acceptable. As a result, university graduates with BS and even MS degrees are forced to apply for less qualified jobs than their NWCG counterparts who have gained experience and training since high school. The authors present recommendations on how to move forward. Most notably is the suggestion to adopt the Association of Fire Ecology accreditation to guide the establishment of new BS degrees.

In thinking to the future, it is clear that the *Journal of Forestry* needs to continue to encourage free thinkers, otherwise termed rabble-rousers, to submit opinion and perspective articles that are contrary to the "accepted status quo." In recent years we have seen a reluctance of journals to accept this type of submission. Too often, however, disciplines become stagnant through only accepting articles that are in line with established opinion. This leads to only incremental advances, if any. However, as highlighted by the articles in this part, it is only through standing up and speaking out against the accepted dogma that advances in any discipline can be made.

Mileposts of Progress in Fire Control and Fire Research

H. T. Gisborne
Senior silviculturist, in charge, Division of Forest Management, Northern
Rocky Mountain Forest and Range Experiment Station, Missoula, Mont.[1]

In this paper Mr. Gisborne gives his selection of the outstanding events in the development of
forest-fire control in the United States. The editor hopes that it will stimulate comments and
suggestions as to other mileposts of progress from many readers of the Journal, particularly
among the "old-timers."

A t Spokane, a month ago, Dr. Carl F. Taeusch of the Bureau of Agricultural Economics told us that for common understanding of any problem it is essential that we first establish the "framework of ideas within which we are trying to work out the problem." Walter B. Pitkin stresses the same point when he says: "The next task is to take some position to ward the subject you propose to discuss.... When you take a position toward a field of fact or fantasy, you do something strangely like the choosing of a vantage point from which you view a valley. . . Point of view and perspective determine the special arrangement and design of the entire panorama. Any valley may be seen from a thousand outlooks. Always the same valley, it presents itself in a thousand manners according to the outlook." Personally, I welcome the

opportunity to develop this topic and to do my part to make sure that we are at least all looking at the same valley and that we are not mistaking some of the minor gulches for the main valley.

I am convinced that failure in recent years to look at the whole problem first, and its parts second, has caused a large part of what we now call the fire problem. For the past 20 to 25 years, we have been so engrossed with certain exceptionally pressing phases of the fire problem that at times we have almost lost our perspective. I believe that I can demonstrate this to you. In the process I hope that I can establish a common vantage point and thereby reveal the true size and complex topography of our "valley." To do this, I propose to go back to the beginning of nation-wide fire control and come down the 35-year road, citing what I have selected

---

1. Presented December 1, 1941, before the U.S. Forest Service Fire Control Research meeting held at Priest River, Idaho, December 1–6, 1941, and edited by A.A. Brown.

as the outstanding accomplishments. In this selection I have had the help of Major Kelley, Axel Lindh, C. K. McHarg, Roy Phillips, W. W. White, and Clarence B. Sutliff. I take full responsibility, however, for all omissions of events which you may think should have been included.

In venturing to select and designate certain events as mileposts of progress, I realize now, better than I did when I started, the difficulty of the task. Probably no one will agree with all of my selections. My major criterion has been whether or not the event was of national and lasting significance in fire control,—federal, state, or private. In some cases I have used the first proposal or broaching of an idea failed to take, or to become nationally effective until clarified, amplified, or given the right push that put it over. In a few cases there has been real progress along broad lines but I have not been able to name the event or date it. I have also probably overlooked some deserving events.

I have tried to distinguish three types of events—(1) progress toward better objectives in fire control, (2) progress through better methods in pursuing the objectives, and (3) progress through more adequate financing which permitted the use of better methods.

1904.—The creation of a widespread framework of Forest Reserves in 1904, 1905, 1906, and 1907, with the assumption by the government for the first time of responsibility for the preserving and protecting great areas of public forest land, is set up as milepost number 1. Previous to this time recognition of the forest fire problem and action to solve it had been purely local in scope and significance. Now with the federal government practicing forestry on a national scale, the nation-wide problems in forest-fire control objectives, methods, and finance began to be defined and the way cleared to deal with them on a similar scale.

This led naturally to the recognition of individual state and private fire-control problems as parts of something that could not be fenced in by land ownership boundaries and dealt with entirely on that basis in *either* the public or private interest. Hence, although Sargent compiled the first fire statistics in 1880, the active history of the fire-control undertaking seems to me to commence in 1904.

1905.—In 1905 was born the first cooperative agreement and work plan between private and state timber protective agencies. This move proved to be so profitable that the formal cooperative protection movement spread rapidly, first throughout the Northwest and subsequently throughout the nation. This first cooperative association greatly improved the objectives and reduced the costs of state and private protection. It was so beneficial that similar "cooperation" has even been forced by law in many states, beginning with the Oregon compulsory patrol law a of 1913 and continuing to date.

1906.—The *Use Book* for 1906 was a definite milepost in bringing all of our major problems of objectives, methods, and finance into focus in terms of the job ahead. "At the beginning of the summer season, or before March 15, each supervisor will recommend to the Forester the number of men needed adequately to protect his reserve, the rate each should be paid, and the number of months each should serve." There, in one pregnant sentence, are all of our problems—beginning, ending, and degree of fire danger, man-power placement, "adequate" protection, temporary employees, and finances. Even Mr. Headley's major problem was there, in the next sentence: "After consideration of these recommendations the Forester will fix the number for the full summer force of each reserve, and this allotment will be final." As Gowen and Headley will probably be willing to testify, the Forester's office is still "considering" similar recommendations and still trying to make allotments that will be "final."

1909.—The formation in 1909 of the Northwest Forest Protection and Conservation Association was more of a national milestone of progress than many of you may think. This organization soon became the Western Forestry and Conservation Association and under the exceptionally aggressive leadership of E. T. Allen, the steady guiding hand of C. S. Chapman, and the farsightedness of Geo. S. Long, it has had a nation-wide influence for more than 25 years. It was especially helpful in promoting federal appropriations for fire control, fire research, and fire-weather forecasting. It exerted and still exerts an extremely beneficial influence on many state legislatures for the improvement of fire laws, brush-disposal requirements, and other objectives. It has contributed steadily to the improvement of methods of fire control, especially on state and private land, through its several editions of The Western Fire Fighters Manual. This text on fire control is probably used in every forestry school in the country. If not, it should be.

1910(a).—This is the first and the last milepost which I can find contributed by a Chief Forester of the United States while he was Chief Forester. If any man here has not read Forest Service Bulletin No. 82, *Protection of Forests from Fire*, by H. S. Graves, he should take time—make time— to do it. There you will find such a keen and dear analysis of the fire problem that many of its statements have not yet been improved upon.

For instance, if you think that methods of fighting fire, crew organizations, skill in sizing up and attacking fires, are new problems, listen to this: "The following are of first importance: (1) quick arrival at the fire; (2) an adequate force; (3) proper equipment; (4) a thorough organization of the fighting crew; and (5) skill in attacking

and fighting fires." What factors have we added to that list in the 30 years that have fled since Graves saw these phases of the problem and described them so clearly? While many old-timers have probably now forgotten this bulletin, many later events indicate that much subsequent progress began right here.

1910(b).—I may be wrong in erecting a monument of progress to the great Idaho fires of 1910, but I am told that before they occurred public interest in forestry and the Forest Service was almost nonexistent. The dramatic incidents and loss of lives in those fires made newspaper headlines all over the country and wakened "the people" to two things. First, that timber wealth was burning up; second, that there was a federal organization trying its best to protect that wealth. A marked up swing in public interest and in funds for fire control was evident immediately after these 1910 fires, which were probably the greatest object lesson as to the importance of fire control that ever occurred, anywhere.

1911(a).—The first forest fire deficiency appropriation in 1911 was certainly a milepost of progress in finance. Some of you may not know that in 1910 several supervisors in Region One furloughed all of their unmarried rangers for one month in order to acquire the funds to pay firefighting bills of the previous summer. Such situations, which were typical of the period prior to 1911, reveal the most serious weakness in organized forest-fire control at that time, inadequate and poorly adapted financing. They give emphasis to the very close relationship between policies in fire-control financing and the progress attained. The provision in 1911 for E. F. F. F., emergency fire-fighting funds, more commonly known as FF, remedied this weakness and made aggressive fire suppression a matter of accepted policy. Most unfortunately somebody made this godsend into a Damocles sword which has been dangling over our heads ever since, Looking back at its origin, it does not seem conceivable that Congress could possibly return to the fire financing methods of 1910, with all that such methods imply.

1911(b).—While the Weeks Law of 1911 was aimed primarily at the acquisition and protection of the headwaters of navigable streams, and hence was a milepost in objectives, it also appropriated federal funds for the first time for fire control in cooperation with the states and there by became a milepost in finance. It was the fore runner of the monumental and highly effective Clarke-McNary Act that followed 13 years later.

1912(a).—*The National Forest Manual* issued some six or more sections, 1911 to 1913, contained in the volume on "Protection" the first detailed breakdown of the fire problem which I have been able to find. There is not time here even to list all the features separately recognized. "The Fire Plan" was named, however, and this was clearly the forerunner of du-Bois' "System atics" for California which followed 2 years later, and for Hornby's highly detailed integrations which followed 12 years after that. Firebreaks were given great emphasis, which, rightly or wrongly, has carried through in some form for 30 years. Permanent lookouts were indicated as possibly desirable as follows: "Main lookouts are those from which an exceptionally large territory can be seen and where it might pay to keep a permanent lookout." This last statement indicates lack of coordination in the Washington Office because the same year that this manual came out, Forest Service Bulletin No. 113 appeared with photographs of permanent lookout towers and houses already in actual use on the Arkansas National Forest. Field practice seems here to have been ahead of Washington Office recommendations.

This manual of 1912 also took a definite step ahead in objectives with the statement that "Practically all of the resources of the national forests are subject to severe injury, or even to entire destruction by fire. Besides the direct damage which fire may do to merchantable timber, to the forage crop, and to watershed cover..." For the first time that I can find, "the forage crop" is included in addition to commercial timber and the old stand-by—watershed cover.

Dr. Shea will be interested to note that 25 years before he entered the field of prevention research, our forefathers stated, "Since the best way to stop fires is to prevent them, a fire plan must include a careful study of prevention methods." Note that they said "prevention methods." I believe you will all agree that we were a long, long time getting past the mere listing of prevention cases and concentrating on the study of prevention methods! There is a vast difference.

Here I would like to digress for one-half minute on the subject of cases versus methods. I have attended quite a few fire meetings, and I have been struck at most of them by the time spent in attempting to solve cases, with so little effort intentionally directed to draw either methods or principles from those cases. Twenty years ago Howard Flint wrote, in a comment on Sparhawk's liability ratings, "Why not stick to a method that is fundamentally sound, using figures that are admittedly arbitrary?" I certainly believe that we, here at Priest River in 1941, should keep our eyes open for methods and avoid quibbling over the split-hair accuracy of minor figures or cases that are, perhaps, being used in an unsound method.

1912(b).—Daniel W. Adams' *Methods and Apparatus for the Prevention and Control of Forest Fires*, Bulletin No. 113 published in 1912, is so clearly the forerunner of both the Fire Control Equipment Handbook and the various fireman's guides and fire-suppression handbooks, that were to follow 20 years later, that it certainly rates a monument in methods. Yet I will venture the guess that not more than two men here ever heard of D. W. Adams or have any recollection of this bulletin. But if you doubt that this was a "first" and should be recognized as such, look at the drawing, in Figure 8, showing where to locate a fire line on a ridge, and compare it with the drawing for problem 6, page 21, in the Region Five *Fire Control Handbook* issued in 1937; or with problem 1, page 88, of the Region One *Fireman's Guide* issued in 1940; or with Bob Munro's Figure 4 in his article in the October, 1940, *Fire Control Notes*. That old drawing of 1912 shows not only the best fire-line location, but also the convection currents involved just as well as many of the similar attempts years later.

As for equipment, if you think that the Los Padres shield for a flame thrower, illustrated in the July, 1941, issue of Fire Control Notes, is something new, take a look at Figure 2 of Plate III of Bulletin No. 113 published 29 years ago. For chemicals on the fire line see Figure 2 of that same bulletin. For a quick get-away with water tanks on a pack horse, see Figure 3 of Plate IV. For railroad tank cars see Figures 3, 4, and 5. For something really new, see the logging system suited to better fire control, outlined by Figures 6 and 7.

Incidentally, this pioneer work in fire control in Arkansas seems to have borne fruit. Dean Walter Mulford recently stated that "Arkansas,which has 15 million acres of active forest lands, is probably the foremost state in the United States as regards forestry matters." I believe that we here should salute D. W. Adams and the Arkansas National Forest. They were so far ahead of us in some respects that we have not caught up yet; if we are not careful we may include in our research program a project or two aimed at features of the fire problem that were pretty well thought out as much as 30 years ago.

1914.—In 1914 Coert duBois' *Systematic Fire Protection in the California Forests*, an unnumbered publication that is not labeled as either bulletin or circular and is marked "Not for public distribution," was very definitely a national milepost in progressive thinking on methods even though the Californians did try to keep it exclusively to themselves. I read that bulletin from cover to cover several times when I was a look-out in 1915. It was all new to me then. Every time I read it now I still find something that is new and useful.

DuBois pointed the way for nearly everything that Hornby and I ever did when he said, "A way must be devised of reducing all of these factors (inflammability, season, risk, controllability, liability, and safety) to concrete terms, so that any forest area, after careful study, can be given a rating which will convey to our minds something in the nature of an exact measure of its total fire danger." The expression "class 5.8 danger in a high-high fuel type" does that for Region One men today, for any instant and spot, with one exception— "liability" or values. They are still omitted. Hornby wanted to include those in his "total danger rating" and they were in his first formula, but the 10 a.m. policy came along about then and under it "values are out." Some day we will go the rest of the way for Coert duBois and put those "liabilities" or values back into the prominent place they deserve, but there is one small matter which must be settled first, the subject of "objectives in fire control." When we clear that up the road will be open again.

1916.—It may have been a coincidence, but if it was a monumental one, when in 1916 Silsocox first proposed the one-tenth of one per cent objective of fire control and about the same time Headley proposed the "least cost plus damage" or "economic" objective. To me, the flat "tenth of one per cent" was an expedient, only a little bit more sound than the 10 a.m. policy which was to come 19 years later, while Headley's theory was and is fundamentally the soundest yet proposed. It is difficult to apply properly, but if we can approach it in the way which duBois believed essential and which has been applied so successfully to danger ratings by "reducing all of these factors to concrete terms," we can make that theory work. When we do that we shall really have applied economics.

1919.—There is some evidence that the Canadians were ahead of us in the use of airplanes in fire control, since this phase of progress did not develop until 1924 or 1925. But Howard Flint had long been investigating lighter-than-air craft, and by 1920 the U. S. Army had become interested. The latter is witnessed by Erle Kauffman who, in an article in the April, 1930, issue of American Forests, quotes an army officer (whom he should have named) as follows: "The day will come when large numbers of men and equipment will be carried by airship to the scene of a forest fire, both men and equipment dropped by parachute, while the airship will rain down fire-extinguishing chemicals from above." From this use of the term "airship" it appears that this army officer was, like Flint, thinking primarily of dirigibles.

The earliest printed record of Flint's interest, which I can find, is in an issue of The Forest Patrolman (Western Forestry & Conservation Assoc.),[2] which quotes Major Kelley, then fire inspector out of the Washington Office, as follows: "H. R. Flint, fire chief in District One, holds credit as the first forest officer to recognize the possibility of the real value in the dirigible as a vehicle for transporting fire crews and supplies, and as a means of effective patrol and detection service. In the fall of 1919 Flint corresponded with a concern in the East about the use of lighter-than-air machine." Flint,

himself, in the December 7, 1931 issue of the *Northern Region News*, says that airplanes were first used on fire control work in this region in 1925. Possibly first use of planes in other regions antedated this. I hope that someone will check that point.

There are three features of this development well worth noting. First, the long, slow, uphill drag indicated by part of Flint's News note: "In the seven seasons since that time, backed up by a little real support, but accompanied by a great deal of discouragement, and some ridicule, I have seen the airplane slowly taking a definite place in our work. It has come to stay." Second, an entirely unforeseen value of this new departure almost usurped the place of the original idea. Photographic mapping, pioneered in 1925 by Flint, J. B. Yule, and T. W. Norcross, almost stole the show and for several years was more significant nationally than was the fire-control idea. Third, Flint's original idea involved the use of blimps. That has not yet come to pass. But the blimp idea obviously led to airplanes, and it is not at all inconceivable that the latest development in airplane use—parachuting men and supplies—may later lead back to the blimps.

1920.— I believe that all fire chiefs, fire bosses, and rangers, will agree that when Orrin Bradeen a began in 1920 to centralize the purchasing, packaging, and delivery of fire-fighting food, tools, and other equipment, he removed one of the great headaches of previous fire-control efforts and made the future job both more efficient and less costly. Bradeen erected a milepost in methods and finance from which we have forged ahead, probably to as near perfection as in any phase of our problem.

1921.—Radio, like airplanes, also opened a new epoch in fire-control methods. While the Radio Laboratory of the Forest Service has, without doubt, made steady progress,

---

2.　Made available to the writer through the courtesy of Mrs. H.R. Flint.

credit for the milepost should go to R. B. Adams, who first made radio actually work on a going fire in 1921, and to Dwight Beatty, who 6 or 7 years later produced the first truly Forest Service sending and receiving set.

1922.—Some time in the early twenties, a new idea evolved which has since become standard practice in all Forest Service regions. This is organized, nation-wide training. While it began as general administrative training, its value in fire control was soon appreciated, and fire-training schools and correspondence courses are now recognized as indispensable. The one man who deserves most credit for this milepost of progress is Peter Keplinger. His full contribution amounted to much more than the first idea, for he stayed with it and developed the method, showing all of us how to use it.

Fred Winn informs the writer that on November 26, 1916, District Forester Paul G. Redington sent an official message 40 miles by radio and then by wire to all other regional foresters and the Washington office. The radio used was built by Ranger Wm. R. Warner and Ray M. Potter, radio amateur. A portion of the pasture fence was "borrowed" for use as the antenna.

1923(a).—Up to 1923 I cannot find a single event produced by Research that should be called a milepost of national progress in fire control. Mr. Clapp's first working plan for fire research, written about 1916, and the research work of Sparhawk, Show, Larsen, Hofmann, and Osborne from 1916 on would rate a tremendous monument in the history of fire research alone, but fire research is only one means to an end and here we are discussing the big end.

Publication of the relative humidity theory in 1923 by J. V. Hofmann and W. B. Osborne seems to me to be the first contribution by research which was of nation-wide significance. The "relative humidity" idea

literally and actually swept the country. For a while it appeared to be the total and final answer to the 1916 Working Plan request for "some simple, single index" of fire danger. For certain individual fuels it is still the simplest and best.

There is one feature of this milepost which should be of special significance to this Priest River assembly of fire-control and fire-research men. While Hofmann was an experiment station director engaged in full time research, Osborne was chief of fire control in this region, and an administrator. But there was no "fence" here between Research and Administration, and this happy combination of an investigator and an administrator proved to be highly efficient. Brevier Show, then a full time investigator, and Ed Kotok, another fire chief, had also joined forces about that same time, and the world knows what a prolific source of ideas, methods, and techniques that partnership became. It continued to be just as efficient after these two men swapped jobs in 1926, when Show became regional forester and Kotok director of the California Forest Experiment Station.

1923(b).—When Show and Kotok, in 1923, distinguished between the "economic" and the "minimum damage" theories, I believe that they erected a milepost which should have accelerated progress more than many of their later contributions. On page 59 of Circular No. 243, they demonstrate that, even while proposing the minimum damage theory, they also favored the hypothesis of least cost plus damage. For they state, in their summary: "Successful protection is reached at the point where the cost of prevention, suppression, and damage is a minimum." Hence, "minimum damage" was offered not in opposition to the "economic theory" but merely as a brake to unsound applications. This is clearly evident when they state, on page 4, that their main objection to the economic theory is that it will not work when too much emphasis is given to holding down the costs of prevention and presuppression. My own opinion, based on 19 years of observation in Region One, supports this view. The economic theory is fundamentally sound. Its defects arise from the manner of application.

1924(a).—The Clarke-McNary Act of 1924 hardly needs any justification as a milepost of national progress. It recognized more clearly than did the Weeks Law the federal interest, hence responsibility, in fire control on both private and state forest lands, and it provided those highly essential funds without which the best ideas must lie dormant. It did another significant thing in reviving a phrase from the *Use Book* of 1906, "to adequately protect." That was the stated objective of fire control on the national forests in 1906. It is repeated as the objective of federal, state, and private cooperative fire protection under the Clarke-McNary Law in 1924. But, I ask you in 1941, what does it mean "to adequately protect," "to provide adequate protection"? I am quite sure that we do not know specifically what we are talking about when we put these words into our federal laws and fire control manuals. We should.

l924(b).— The first written agreements between the Weather Bureau and the Forest Service providing for "fire-weather warnings" were dated August 11, 1916, and March 12, 1917. I cannot regard them as milestones of progress, since they provided for measurements of wind velocity only from the forest stations, and the forecasts, when furnished, were of doubtful value. In about 1924, however, a meeting of Weather Bureau and forest protective agencies was called by the western Forestry and Conservation Association, at which methods of measurement and types of forecasts were thoroughly discussed. Out of this

came the first congressional appropriation of funds specifically for fire-weather forecasting. I believe that this meeting in 1924 rates the mile post.

1925.—Everyone here is aware of the vital and extremely practical problem of allotting fire-control funds. You fire chiefs probably appreciate it most keenly. It cuts you the most. Headley and Gowen undoubtedly know more about it than anyone else in the world. How many of you know that Sparhawk worked on this particular problem from 1915 to 1921, and wrote a bulletin on it that was published in 1925?[3] In his tremendous compilation and analysis of Forest Service fire records from all over the country, Sharhawk was not attempting to tell anyone how much money should he allotted to each region, forest, and ranger district. He was hunting for a method by which that could be done, and, of more significance, for an over-all justification for fire control expenses. He was trying to answer that extremely basic question,—what is the cost of adequate protection?

Sparhawk had to work with inadequate and inaccurate fire statistics. He was not in as good position as we are today to define adequate protection, to formulate correct economic balance, or to develop economic justifications. Yet we can claim little advance in these respects over the clear analyses he developed in 1925.

As evidence of his attitude toward the economic theory, Figure 1 of his report carried a diagram illustrating the effect of the law of diminishing returns. It is the same diagram Mr. Headley was using last summer in an attempt to re-awaken forest officers to the full implications of this relationship. The term "hour control" was used by Sparhawk and was the starting point of work such as done by Show and Kotok in their "Determination of Hour Control." But

3. Sparhawk, W.N. The use of liability ratings in planning forest fire protection. Jour. Agric. Research 30:693–762. 1925.

they avoided the question of justifiable cost of meeting a particular hour-control standard, and such standards gradually became an end in themselves. Their only justification is as a means of attaining adequate fire control, but adequate fire control still lacks any exact definition.

Sparhawk knew the necessity of answering such questions. He listed, for the first time to my knowledge, all the kinds or values of forest resources which justify protection. No one has since added anything to that list of timber, including mature timber, young growth, the forest capital, and soil productivity; forage for live stock; regulation of stream flow and the prevention of erosion and floods; game resources; recreational use; improvements; and other occupancy values. That list is a true masterpiece of perception. Note, for instance, the inclusion of "the forest capital." When the silviculturists get a working circle into managed age classes, it is obvious that a fire which destroys or even seriously depletes one or two classes does damage far exceeding the maturity value of those particular trees discounted to date. The whole working circle is thrown seriously out of orderly future progression, and a form of loss results that is still far ahead of us in 1941. Sparhawk saw it, and named it, in 1925! Research program makers of 1941 can well go back and begin at this 1925 milepost n many respects.

1926.—This milepost, *A National Program of Forest Research* by Earle H. Clapp, is so well known and so well appreciated that it needs no explanation. There are certain features of the forest protection section, however, which I should like to emphasize.

First, protection from fire, protection from fungi, and protection from insects are tied together so closely that every time I read these three sections again 1 wonder why the Forest Service has a solitary division of fire control when the job, on the ground, could be so much more efficiently handled by a division of forest protection. If you will follow this through, you will find that here may lie one of the most effective methods of solving the problem of the temporary employee, or keeping a trained organization.

Second, is another case of lack of coordination in the Washington Office. In 1925 Sparhawk named all of the many reasons for fire control, of which commercial timber was only one. A year later, Mr. Clapp implied, by his words "forest management" and "to grow timber," that timber alone for dollars alone is our major and perhaps sole justification. Are we still at odds on this issue, or is there now a closer coordination of ideas?

Third, that Mr. Clapp subscribed to the economic theory is shown by his statement: "Possibly they [foresters] should also be able to set limits beyond which expenditures for protection are not justifiable, that is, the determination of that point where the law of diminishing returns becomes effective." In his next sentence he opened the door to the ultimate answer when he said, "But if used at all these limits should be set upon very comprehensive rather than narrow considerations." As will be brought out later, the dollar value of destructible resources is not the sole criterion of damage, and a much more comprehensive basis, as Mr. Clapp called it, is absolutely essential.

Fourth, under "Protection Standards," Mr. Clapp stated, 15 years ago: "Satisfactory timber crops cannot be grown unless certain definite standards or objectives of protection are attained. . . . Those standards must be definitely determined.... This is a task best attempted by research methods."

1928(a).—The McSweeney-McNary Act of 1928 is perhaps merely a result of the 1926 national program, but it is well to point out one difference. The "program" was an idea, a plan for a "functional operation." It could not function, however, without finances. The McSweeney McNary Act liberated these essential dollars, like putting water into an irrigation ditch. A lot of fine work had gone into building the ditch and laying out the orchard, but until some water was turned into the ditch the orchard could be neither planted nor irrigated. The McSweeney McNary Act permitted both plantation and growth.

Here, again, when we come to our fire research program, let us remember our history. We here at Priest River are merely extending that same ditch and laying out some more orchard. The other half of the job still remains to be done. Somebody must divert water into our ditch. Unless that job is specifically assigned, and another McSweeney-McNary Act puts golden water into our ditch, our work here at Priest River will join Sparhawk's fine work on the bookshelves.

1928(b).—Another milestone in methods was erected in 1928, or thereabouts, which should be credited specifically to the chief of our division of Engineering, T. W. Norcross. On the basis of certain personal information I date this milepost as 1928, instead of 1931 when the Norcross-Grefe report was published. When Norcross saw this opportunity, devised a systematic attack, and rang the bell with his transportation planning methods, he gave all future investigators and administrators a well-designed tool. In my opinion that is a major contribution.

1930(a).—Although Sparhawk may have originated the "hour control" idea in the early twenties, and Norcross designed transportation planning to meet hour-control standards in the middle twenties, there is no doubt that Show and Kotok erected a milestone in 1930 when they published Technical Bulletin No. 209, *The Determination of Hour Control for Adequate*

Fire Protection in the Major Cover Types of the California Pine Region. This popularized the idea and the term sufficiently to produce action in many parts of the country.

In this bulletin, Show and Kotok also added to all previous concepts of "adequate protection"when they went one step beyond Silcox by setting "an annual average of 0.2 per cent for the commercial and potential timber types and at 0.5 per cent for the nontimbered types" as the criteria of adequate protection in their region. Here was recognition of something new, an objective varying according to economic demands. There were many other outstanding features of this particular publication, but none, to my mind, either here or in their earlier *Role of Fire and Cover Type* bulletins, which so strongly influenced national ideas and action as this variable standard.

1930(b).—The milepost erected by the District Foresters' Washington meeting in 1930 seems to have been largely a nation-wide application of the "variable standard" originated by Show and Kotok. In view of all previous statements of fire-control objectives, it is well to note here how the district foresters affirmed the past basis of damage as a criterion: "Damage from fires to forest values varies considerably in the different forest types and the objectives in fire control must be based mainly upon consideration of these variations in damage." The fire control committee, of which Kotok was chairman, then listed (1) timber, (2) site, (3) reestablishment process, and (4) future .fire danger; as the four main features of damage. Forage, recreation, and game were ignored, as well as that feature which Show and Kotok were to inject later, "downstream federal financial interests." It is evident that while the objectives were incompletely stated, still there was no doubt that damage was the sound basis. Every milepost in objectives from 1906 up to, but not including, 1935 will be found to be in agreement on that point.

1933.—The advent of the C. C. C.'s in 1933 appears to me to have been a milepost first in finance and second in methods. Money and labor of a sort were here made available to carry out Norcross' transportation plans, to approach Show's and Kotok's standards of speed of attack, and to build more airplane landing fields for Howard Flint. Besides implementing these ideas, the C. C. C. program gave urgency and immediate application to the development of new systematic methods of presuppression planning, involving new concepts in transportation systems, detection systems, and communication systems. The planners who charted out the methods and developed the techniques had, for the first time, the satisfaction of seeing them tested and applied on an adequate scale in an unprecedented investment program. This area of presuppression planning is still too recent to evaluate fully. A large part of the fire-control group present have taken leadership in some or all phases of it. A whole body of literature has grown up around it, with many new concepts that are finding increasing application. Cooperation between federal, state, and private agencies also was pushed ahead in a significant surge. Coincident were funds for considerably expanded research of certain types, and the construction of better plant facilities.

The C. C. C.'s brought a surge in methods too, or with them came both the opportunity and the necessity of training fire-fighting crews. Development or improvement of the man-passing-man, the sector, the one-lick, and the 40-man shock crew methods of large-fire suppression all seem to have been accelerated. New ideas not only grow best but can be tested best when both money and men are available. A similar period seems highly probable after the end of the present war.

1934.—I have been advised by some of the consultants who helped me prepare this list of mileposts, that fire danger meters should be included. Because of my personal interest in those particular devices, I am automatically disqualified from judging that point. I therefore leave them out. [It would be unfair to let this statement by Gisborne stand without positively adding a milepost to the "fire danger meter" which he introduced in 1934. This was actually doing something about the philosophy for which he eulogizes Coert duBois. He, a "researcher," undismayed by the lack of basic studies, boldly proceeded by empirical methods to combine the influence of weather factors, season, and fire risk at any given time into a single rating which the confused administrator could use to determine the intensity of fire organization called for on the ground by each combination of factors. This was a new and better answer to an old idea. Gisborne's fire classes, and the meter by which they are determined, made possible for the first time a simple expression of exceedingly complex relationships. His idea has since been applied throughout the country, with the many variations and adaptations requited by local conditions and needs, and has given rise to a bewildering array of fire-danger meters, each with some new and valuable feature. It deserves a major monument not only for its contribution to more efficient fire control management but for the part it has played in reorienting fire-research programs and in giving new stimulus and significance to investigation of relationships and factors not yet clearly understood.—A. A. BROWN.]

1935.—The so-called "Forester's policy of control by 10 a.m." undoubtedly rates either a milepost or a tombstone on our 35-year road of progress. If and when that policy becomes clearly recognized as a temporary expedient, I believe that it will rate a milepost. If, however, it has already become or ever does become the death knell of all previous objectives based on damage, then it rates a tombstone executed in the blackest of black granite. It has already cost us 6 years of neglect of variable damage as an objective, but it seems to have achieved something else which may have been, at the time, worth more than the little thought which might have been given to damage.

It is futile to open a discussion of that policy here and now. It has such a direct bearing, however, on any fire-research program which we may recommend that its import deserves serious thought. First of all, it is important to recognize that while the 10 a.m. policy theoretically specifies the same standard of protection for commercial timber, reproduction, forage, water control, recreation, and wildlife, in practice it is not fully enforced. To this amateur historian, it appears that the policy actually had the same objective as the Show and Kotok minimum-damage theory of 1923, to wit: Stronger prevention and presuppression action to catch fires small, rather than stronger suppression action aimed primarily at keeping 10,000-acre fires from becoming 20,000- or 30,000-acre burns. There is a vital and basic difference here which will come out in our discussion of the economics of fire control. But if the main idea of the policy was to control fires while really small, the use of a time criterion would seem to be open to further investigation; for fires can be caught small and cheaply, often more cheaply, without controlling them by 10 a.m. tomorrow. If one function of research is to assemble and array all the significant facts, it seems more than possible that it might contribute something here.

1936.—Hornby's methodical treatment of all the significant features of fire control, especially this weighting of each factor and final integration of all of them, has been approved as a milepost of progress by all of my advisers. Perhaps his most outstanding contribution was his analytical approach to the planning problem and his re-emphasis on physical conditions on the ground as the proper starting point for all fire plans. His concentration on fuel types, rate of fire occurrence, rate of spread, and fire danger, as fundamental, measurable factors of the fire job everywhere, constitutes a sound basis for future progress. Fire-control planning work is not new. It began with the organization of the Forest Service. But Hornby systematized that planning, made it so methodical, and incorporated so many new features that all future fire-control planners were greatly aided.

1938.—From Hornby's milepost in 1936 to date I cannot find a single event in objectives, methods, or finance that has proved to be of national significance in fire control. I believe that there are three reasons for this. First, in any field of human endeavor, whether it be forest-fire control or the effort to produce a temperature of absolute zero, the nearer you approach your goal the harder it is to take the next step ahead. Second, progress requires men and time to work, and these require funds. Since 1936 funds for all kinds of Forest Service work have been steadily reduced. While we have had additional "relief labor," intellectual progress in objectives and methods of fire control has not been and never will be assisted in the least by E.R.A. and W.P.A. labor. Third, it is difficult to judge the most recent past. Perhaps some steps have been proposed or taken during the last 5 years which will show up later as milestones of progress.

One recent step of which I believe this may be true is illustrated in Region One by the scheme devised by Sutliff in 1938 for maintaining a standard relationship between current fire danger and the percentage of man power on duty. Shank had established a standard relationship and was using it on Region Four forests shortly before this, and Brown started his "Step-up" plan in Region Two in 1937, all of which were moves toward this one objective. The acceptance of this scheme in Region One has for 4 years done for current fire danger exactly what Hornby's systematic planning did for average bad danger. Hornby's method says that when the permanent factors of danger are thus and so, the following list of stations and facilities must be available for occupancy and use. Sutliff's table X-1-c says that when the variable factors of danger are thus and so, the following percentage of those stations will be occupied. These are two clear-cut, logical steps, both essential to adequate fire control at least possible cost. In 1958, when all forests have planned alike, when all are provided with facilities according to uniform consideration of the same factors, and when all are manned alike according to uniformly measured danger, the principle represented by these methods of correlating man power and fire danger may be judged as a milepost of progress. The possibility is at least sufficiently great to justify its mention.

While this concludes the list of definite and datable events which I rate, now, as milestones of progress, there has been one other type of steady progress that must not be ignored. This type is difficult to name and impossible to date. It is illustrated by Kelley's Fire Code for the eastern forest region issued in 1926, and by Headley's *Fire Control as an Executive Problem*, mimeographed in 1928. It is illustrated by the drive to "calculate the probabilities" contained in the Forester's 10 a.m. policy of 1935 (which is the best part of that policy) . It is the concept that fire control is a tremendously complicated job, but one which is susceptible to orderly dispatch if the man uses his head, looks at all the factors and facilities, forms correct conclusions, and then takes action. To me it seems that this particular phase of the fire problem began when the Forest Reserves were created in 1904, has become increasingly important since then, and will never end.

I am purposefully not saying anything about "regulation" as a milepost in the progress of fire control, although that idea, and especially the recent action concerning it, are certain to exert a tremendous effect in the future.

Conclusion

If this summary of some of the developments of fire control appears to be primarily an attempt to pass judgment on history, then I have failed in my main purpose. My real purpose was to collect, select, and relate enough of the major events of fire control during the past 35 years so that we would have a reasonably dependable background or stage for this Priest River meeting. It may be significant that 19 of the mileposts mark progress in methods, 14 are achievements in understanding our own objectives, while only 11 major steps are evident in finances. One might question whether this shows knowledge ahead of practice, or finance retarding application.

I have tried also to assemble the "framework of ideas" within which we are trying to work out the problem, to see the problem as a whole in the light of past accomplishments. My viewpoint is naturally influenced and perhaps controlled by my own personal experiences to date. I cannot help that, but I admit that mine is no the only viewpoint. Others undoubtedly see this field of fact from a different viewpoint. I know that I should benefit by having them do for me this same job that I have tried to do for them: review the field and tell me how it looks to them. We have done altogether too little of that in the past 20 years and our failure has constituted a serious weakness in our work. I am convinced that many of our present disagreements would dis appear if we could get together and look together at the whole valley from each of the several admittedly different vista points. and I am convinced that we here at Priest River cannot expect to lay out a sound fire-research program unless we keep in mind the major events which constitute the history of our particular line of work.

Comments

Roy Headley, *U. S. Forest Service*
It is not surprising that Gisborne has a few of his facts wrong. The surprising thing is that he found time to get so many of them right. This business of identifying mileposts in American forest-fire control takes hard, careful digging.

Silcox selected a lookout station on Squaw Peak on the Cabinet National Forest in 1907. The idea dates back that far at least. But who first thought out that method of detection? and who first actually put the idea into effect? I never asked Silcox. If we knew the answer it might help us to set the stage for equally important inventions or adaptations.

The use of aircraft in fire control probably began in California in 1919 when duBois got the Army Air Corps to organize an elaborate aerial patrol system over the national forests in that state. Some of this early history is reported in the article "Wings' and Parachutes over the National Forests" in the October, 1940, issue of *Fire Control Notes.*

Gisborne does not go back far enough for the genesis of training. There were Forest Service ranger training camps at Fort Valley, Ariz., and Hot Springs, Calif., in 1909 or 1910. This form of training then suffered something of an eclipse because of hostile legal control that followed the dismissal of Gifford Pinchot; but it later overcame the legalistic hurdle and became recognized as an indispensable feature of personnel management.

Fire-guard training started (naturally?) in California about 1913. I believe that D. P. Godwin, then supervisor of the Mendocino National Forest, first saw the need and held the first training camps for fire guards. I

doubt if fire training arose out of general administrative training. It may have been the reverse. Men were usually willing to admit that they needed training in fire control before, they could accept the view that they needed training in "administration."

Since Gisborne is the father of danger rating and danger meters, he is modest in speaking of that milepost. In my opinion the danger meter and the ideas and possibilities which go with it are of very great importance indeed. I wish we could have more such fine pieces of fire-control thinking.

In speaking of Hornby's work, Gisborne emphasizes the importance of the "approach." He says it often determines the result of research before the research is begun. If I could, I would double and redouble all that Gisborne says and urge that research of the "approach" should not be neglected as it too often is.

However, Gisborne then argues that "physical conditions on the ground" is the approach to use as against the "results attained with past facilities and funds," in deciding on distribution of funds and facilities for fire control. But where does this leave us with respect to the minimum cost and damage objective which he regards as "fundamentally the most sound ever proposed"? If two sides of a barren-topped mountain range have identical cover and other physical conditions, would the physical-conditions approach lead to identical allotments of funds for fire control? Would it still do this if on one side of the range there is an unused desert and on the other a highly productive irrigated valley subject to water shortage and great flood damage?

And what would we do if according to the physical-conditions approach areas A and B should have identical funds while according to the "results attained with past facilities and funds" approach, area A should have 25 per cent more facilities and funds than area B? Should the experimental values determined by past results be set aside in favor of a man-made rating of physical conditions? If so, isn't that a bit hard on the experimental method as against the observational method? and what shall we think if we find that the "physical conditions on the ground" approach, tends to call for more money than the other approach?

Gisborne hardly expects any one to agree with him on his selection of mileposts. Agreement at the outset is not so important as critical study, which if kept alive must eventually lead to substantial agreement.

Gisborne omits mention of the mechanization of fire control. There may be no event to identify as a milepost, but it has been a most important development which may prove a life saver during the next few years when labor will be so scarce that it will be impossible to fall back on the principle of "fifty men to do ten men's work."

He makes no mention of cooperation in prevention and suppression of the type that John McLaren and others developed. Such cooperation relapsed under the impact of the C.C.C. but is being revived now that the C.C.C. is folding up and idle labor is no longer available for our sudden heavy demands for suppression labor. Local cooperation can mean as much to us as it meant to our army in the Philippines.

While Gisborne mentions the one-lick method he doesn't really give it a milepost rating. The story is important as a study in how ideas are born, languish, and sometimes revive. Godwin was the inventor in about 1915. But the brain child was stillborn and no amount of effort could seem to revive it—until 1935. In that year it was hopefully attached as a rider to a letter on the Chief's 10 a.m. fire-control policy. Sure enough, some one sparked. Kenneth P. McReynolds went into action. See his story in the December, 1936, *Fire Control Notes*. His activity enabled the one-lick method to get going at long last, and must

have been partly responsible for the whole brood of similar methods which have since been hatched. But the important question is, what mistake of leadership was it that kept the one-lick and other ideas for speeding up production of held line dormant for twenty years,—and to that extent retarded an aggressive attack on the problem of the tragically low output of held line per man hour?

Why no milepost for the Mather Field conference in the early twenties? A good many important things started there, including action on the idea of fire-control roads on specifications just sufficient to serve the purpose. A large mileage of such roads actually came into use before that approach to the problem of investment in fire roads lost out.

How about a milepost for the employment of professional psychologist on fire prevention ina1938? Even if we have no statistics on the fires he has prevented, it represented a break in the tradition that, in fire control, foresters need no help from other professional disciplines. That break could lead by easy stages to the use of specialists in sociology, economics, and eventually even in organization and management.

Along with the employment of a professional psychologist went the employment of an expert in obtaining cooperation in fire prevention from national offices of national· organizations. Results in terms of contributed money cost of our series of nationally distributed posters may be spectacular as well as some things national organizations have done, but on a milepost jury I would expect few to vote with me for a favorable verdict.

In 1919 I acquired an acute dislike for the word "smokechaser" because it seemed that the role of the smokechaser was to chase out, look at the smoke, chase back, and wait for a crew from the nearest labor center a hundred miles or more away. Today, it is my belief that our typical smokechasers or firemen perform marvelous exploits of tenacity, devotion to the job, and intelligent heroism. If there is any truth in the contrast, the change surely rates a milepost of some kind. But where? It is the result of years of careful selection, good inspection, and inspirational discipline by many fire-control managers. It is a sample of many developments which, while of vital importance, cannot be identified with a monument or a mile post to mark some particular event.

Why not a milepost for the 1940 start on appraisal of the so-called intangible damage from fire? Tangled thinking had postponed such a start for a long time, but at last it got under way. We could easily get tangled again and fail to follow up on the start that has been made; but it is also true that this beginning could be followed by far-reaching gains in the intelligence with which fire control is managed.

While Gisborne had a special purpose in preparing his paper, we owe him a vote of gratitude for the start he has made on the history of ideas in fire control. As he well says, our future thinking will be more fruitful if we pay some attention to the ideological steps we have taken in getting to where we are now.

For seven years ending with 1941 the national forests of the six western Regions as a whole have more than reached the once seemingly unattainable goal of an average annual area loss of one tenth of 1 per cent. Perhaps there is a "bad" year around the corner which will humiliate us once more; but for the moment, the record stands.

For 35 years everybody has been too busy trying things, making mistakes, and learning from them to write this saga of American forest-fire control. But the time has now come to study its mileposts; also the slow growths of competence in which mileposts are almost impossible to identify. and Gisborne has given us a good send-off.

Fire as an Ecological and Silvicultural Factor in the Ponderosa Pine Region of the Pacific Slope

1943. *Journal of Forestry*. Vol 41. No 1. 7–15.

Harold Weaver

Forest supervisor, Colville Indian Reservation, Nespelem, Washington[1]

This article presents evidence in support of the author's belief that complete prevention of forest fires in the ponderosa-pine region of the Pacific Slope has certain undesirable ecological and silvicultural effects. He emphasizes the fact that conditions are already deplorable and are becoming increasingly serious over large areas, and urges intensive research on the problem.

Wherever man goes in the ponderosa pine region of the Pacific Slope[2] he sees evidence of past forest fires. The occasional charred remnants of old trunks and stumps, the partially burned snags and windfalls of trees killed by various causes in more recent times, and the charred bark and the basal fire scars or "cat faces" on the trunks of numerous standing live trees are everywhere in evidence. Many apparently unblemished trees when felled are found to have hidden, grown-over scars caused by fires that burned at various times during the earlier life of the tree, some as early as the sapling and pole stages.

The writer believes that these facts indicate clearly that periodic fires, in .combination frequently with pine-beetle attacks, and occasionally with other agencies, formerly operated to control the density, age classes, and composition of the ponderosa-pine stands. Undoubtedly the results would often have appeared wasteful and harmful from civilized man's viewpoint.

---

1. Forest supervisor, Colville Indian Reservation, Nespelem, Wash. He writes from a background of 17 years' varied experience on the national forests and Indian reservations of the Pacific Coast. This article represents the author's views only and is not to be regarded in any way as an expression of the attitude of the Indian Service on the subject discussed.
2. California, Oregon, Washington, northern Idaho, and western Montana.

So, in fact, they did appear .in many localities subsequent to the settlement of the Pacific Coast States by the white man. The early forest conservationists consequently concluded that total exclusion of fire was a vitally necessary prerequisite to forest protection and sustained-yield forest management.

As a result of this policy fire has actually been excluded from large areas for from 30 to 40 years. This has brought about changes in ecological conditions which were not fully anticipated, and some of which seem to threaten sound management and protection of ponderosa-pine forests. After considerable study, the writer has concluded that progress in converting the virgin forest to a managed one depends on either replacing fire as a natural silvicultural agent or using it as a silvicultural tool. He can see but little evidence of success in solving the first of these problems, and believes that far too little thought and research has been applied to the second. In support of these conclusions he offers the following evidence.

The Record of the Tree Rings

The precise years during which past fires have occurred can be determined by a study of the rings that show the age of the tree and record the vicissitudes of its life. When fire scars are formed the tree immediately attempts to cover them with new wood, which grows from the living edges of the cambium layer towards the center of the wound. By counting back to where the calluses occur it is possible to determine with considerable accuracy the dates of the fires.

Several intensive tree-ring studies have been conducted in the ponderosa-pine region. Keen, in his study of climatic cycles in eastern Oregon (2), found that fires swept the Watkins Butte area of the Deschutes National Forest during the years 1824, 1838, 1843, 1863, 1883, and 1888. The tree rings also indicated similar fire frequencies for other centuries. On another area he found that one of the oldest trees, dating back to the year 1255 A.D., was originally "cat faced" by a fire in 1481. Subsequently, until the tree was cut in 1936, it weathered 25 fires which occurred at approximately 18-year intervals (3).

Show and Kotok in their study of the role of fire in the California pine forests concluded (6), from their own observations and from a tree-ring study previously conducted by Dr. J. S. Boyce, that extensive fires had occurred in 25 clearly marked years during the preceding three centuries. They also mentioned Huntington's investigation of the giant sequoias (Sequoia gigantea Lindl.), which indicates that in the restricted localities of the pine region where these trees now grow fires occurred as far back as the year 245 A.D.

From the evidence presented by the tree-ring studies it is obvious that fires have occurred throughout the ponderosa-pine region down through the ages. Over most of the region it is probable that extensive surface fires burned just as frequently as inflammable dead needles, twigs and other debris, and dried grass and brush accumulated in sufficient quantities to support combustion and to carry the fire along the forest floor. Fires then occurred whenever weather conditions were favorable and some natural or human agent caused them to be started. One of the most fruitful sources of fire has always been lightning and it is probable that the Indians caused many fires, either intentionally or by accident.

Fire as a Cause of Even-Aged Grouping of Pines

The greatest portion of the ponderosa-pine forests is characterized by the existence of uneven aged stands made up of even-aged groups in various stages of maturity. There is abundant evidence that this condition was caused by frequent surface fires operating in conjunction with periodic epidemic attacks of the western pine beetle (*Dendroctonusb revicomis* Lec.), with occasional attacks of other tree-killing insects, and with windthrows.

It is believed that the process was developed and maintained approximately as follows:

1. A single tree or a group of trees, sometimes a very large group, was killed by attacks of the western pine beetle. Similar destruction probably was caused on Occasion by the attacks of other insects, or by windthrow, and the surface fires themselves occasionally weakened and destroyed large trees by the repeated burning out of large basal fire scars.

2. In the opening or openings thus created ponderosa-pine seedlings germinated. Their subsequent rapid development was encouraged by the comparative freedom from root and crown competition of larger trees.

3. Inevitably another surface fire swept through the forest. The dead snags and windfalls of the large trees that originally occupied the openings were wholly or partially consumed, and many or sometimes all of the seedlings and saplings were destroyed. Usually, however, a number of the dominant, more fortunately situated individuals survived, and these, after recovering from the effects of partial defoliation the fire, put on greatly accelerated growth by reason of the release from competition with their former comrades. If all of the seedlings and saplings in the openings were destroyed the reseeding process was repeated. In some cases the young trees that invariably occupied the openings were subjected to several fires while they were still in the sapling and small-pole stages.

4. After the young trees attained such age and height that they were no longer susceptible to defoliation by surface fires, they were further thinned during the large-pole and "black-jack" stages by the attacks of such insects as the various species of *lps*, the mountain pine beetle (*Dendroctonus monticolae* Hopk.), and the western pine beetle. The dead snags resulting from these thinnings were eventually consumed by surface fires.

5. After the trees attained the thrifty mature, mature, and overmature age classes they were in turn susceptible to destruction by the various agents enumerated in paragraph (1). Thus their life cycle was completed and the space that they had occupied was t ken over by the new generation of trees.

This cyclic process can be seen in operation in certain parts of the forest at the present time. Single trees and groups of trees are constantly being killed by attacks of the pine beetle. In the openings thus created, there can be observed over a period of a few years the actual germination and subsequent development of the ponderosa-pine seedlings. Where recent surface fires of light to moderate intensity have burned there are excellent examples of the thinning process in the patches of reproduction that occupy these openings. Such an example is afforded by the Harm Springs Indian Reservation fire of 1938, on the east slope of the Cascades in central Oregon. On several thousand acres of Tenino Bench the fire did such an excellent job of cleaning up the beetle-killed snags and windfalls and of thinning the adjacent stagnating sapling and pole patches of reproduction that the surviving trees have responded to release from competition and are now making rapid growth.

Throughout the groups of ponderosa-pine poles and "black-jack" groups of age classes in excess of 40 or 50 years can be seen the charred remains of the larger trees that originally occupied the site. By taking increment borings or by chopping to determine the age of the young trees, one can find the approximate date at which the larger trees were killed, and by studying the external and hidden scars which most of the young trees bear, one can determine the exact years of the various thinnings by fire.

Everywhere are the groups of thrifty mature, mature, and overmature trees with their tree-ring record of fires of long ago. The evidence is there for those who care to investigate

Influence of Fire on Even-Aged Stands

Extensive even-aged stands of ponderosa pine can probably be accounted for by the past occurrence of severe crown fires, by severe epidemics of tree-killing insects such as the defoliating white-pine butterfly (*Neophasia menapia* Felder), or by the occurrence of extensive windthrows caused by cyclonic winds. Subsequent to the insect killings or the windthrows, fires of great intensity often occurred.

In the early 1900's the white-pine butterfly killed a major portion of the ponderosa-pine stand on several townships of the Yakima Indian Reservation on the east slope of the Cascades in southern Washington. A dense even-aged stand of ponderosa-pine reproduction now occupies this area. Because of its existence and the dead windfall and snag remains of the original stand, the fire hazard is tremendously increased. Thanks to the protective efforts of the foresters, however, fires have not thus far entered the picture. This stand is now badly stagnated and thinning to get released growth presents a difficult management problem.

Burns at Higher Altitudes

At the higher elevations in the ponderosa-pine region, particularly in the Upper Transition and Lower Canadian zones, where moister conditions usually prevail and the dry seasons are much shorter, fires naturally occurred at less frequent intervals. When fires did occur, however, they usually caused great destruction because of the low-branching growth habits, abundant reproduction, and great amounts of accumulated inflammable debris characteristic of the fir or fir-larch forests that form the climax type on such sites. Lodgepole-pine trees also have always been very susceptible to fire because of their thin bark. Also where such stands have previously been decimated by severe epidemics of the mountain-pine beetle, crown fires of extreme intensity have usually developed

Most of the burns of the Upper Transition and Lower Canadian zones have been restocked to even-aged pure and mixed stands of lodgepole, ponderosa, and western white pines and occasionally, in Washington, Idaho, and Montana, to western larch and Douglas fir.

Importance of Fire as an Ecological Factor

Past history indicates clearly that fire has always been an extremely important ecological factor in the ponderosa-pine region. Dr. Willis Linn Jepson, in his *Trees of California* (I), makes the following statements: "The Sierra Nevada forest, as the white man found it, was clearly the result of periodic or irregular firing continued over many thousands of years. . . . As a result the Sierran forest shows marked reactions to millennial fire conditions. Three observations illustrate this statement. First of all, the forest stands in this belt occur where from the nature of the topography the fire ravage would be expected to be less severe; secondly, the individual trees are extremely

Figure 1. This stagnating stand of 37-year-old reproduction near Nespelem, Washington, is typical of the dense even-aged stands of ponderosa pine that have developed since the advent of total fire exclusion.

Figure 2. A portion of the stand shown in Figure 1 as it appeared in 1941 after being thinned by a surface fire in 1914. In the background and so the right can be seen the edge of the stagnating thicket, along and outside of the 1914 fire line.

Figure 3. This stand near Badge Creek, Warm Springs Indian Reservation, Oregon, is typical of vast areas of the ponderosa-pine region where, as a result of total fire exclusion, dense even-aged stands of Douglas-fir reproduction are monopolizing the ground under the mature ponderosa pines.

well-spaced and commonly form a very open forest, the degree of openness often being in direct ratio to the age of the stand; thirdly, the trees as a whole, without regard to specific relationship, exhibit trunk bark of such unusual thickness and often of such noninflammable character that it may be taken as evidence of protection to the tree if not of marked adaption to fire ravage.

Dr. Jepson then states that the most abundant tree in the Sierran forest is the yellow (ponderosa) pine which has fire-resistant bark 2 to 5 inches thick. In discussing the Sierran forest he says, "Indeed the main silvical features, that is, density, reproductive power and dominance of types, are in great part expressions of the periodic fire status."

Although the fact that fire has always been an important ecological factor is recognized to a certain extent by most foresters, many of them disregard or minimize the possibility of utilizing fire as a silvicultural agent in the management of ponderosa-pine forests. They refer to the past fire history of the stand as "the process of attrition" and assume that the magnificent, open pine forests found by the first white man on entering this region were but understocked, patchy stands, "worn down by the attrition of repeated light fires." In the dense stands of reproduction that have come in as a "veritable blanket" under the mature stands since the advent of fire protection, they seep roof of this interpretation. Show and Kotok (6) for example, state that, "This remarkable change is in itself proof that the virgin forest as we find it does not represent the productive capacity of the land, for if an area of ground is fully occupied bx a mature crop of timber the young individuals cannot obtain a foothold because the available moisture and light are fully utilized."

Present Conditions in the Ponderosa-Pine Forests

From its very inception, the policy-making and the administrative and investigative phases of our fire protection program have been largely guided by foresters of the school of thought which favors total exclusion of fire. As a result. over the greatest portion of the ponderosa-pine region where we have had organized protection, fire has actually been excluded from large areas for from 30 to 40 years. Some of the resulting conditions and changes that the writer considers most significant are as follows:

1. Dense even-aged stands of ponderosa-pine reproduction have developed. Where these stands have established themselves on the better sites, in the natural openings constantly being created in the virgin forests and on the clear-cut and selectively cut lands. they appear to be making excellent growth. Over the greatest portion of the region. however, especially where they comprise "veritable blankets" under the mature trees, they are making exceedingly poor growth. Meyer in his study of growth in selectively cut ponderosa-pine forests (5) makes the following statements: "The slowness with which the reproduction stand develops is one of the most discouraging phases of the management of selectively cut ponderosa pine forests in the Pacific Northwest. . . . Overstocking and clumpiness may be the principal causes: the oft-mentioned poor growth conditions of the last decade or two may be also a factor. On the average, the reproduction stand can be counted upon only to produce about 10 per cent of the normal yields for site index 80. This situation is deplorable. If stagnation is allowed to persist it will imperil

the cuts at the end of the second and subsequent cutting cycles.

2. Enormous areas are growing up to dense, even-aged stands of white-fir, Douglas-fir, and incense-cedar reproduction under the merchantable ponderosa pines. These species are very susceptible to killing by fire, which has largely excluded them from such stands in the past. When fire is kept out, however, they come in under the dense shade of the larger trees, and under such conditions the less tolerant ponderosa pine seedlings have no chance to compete.

3. For the past 20 years epidemics of the western-pine beetle have killed and are continuing to kill billions of board feet of ponderosa pine worth many millions of dollars. These epidemics are apparently more severe than those that formerly occurred in the ponderosa-pine region. In the past two decades, in the states of Oregon and Washington alone, the beetles have killed 15,920,000,000 board feet of merchantable ponderosa pine conservatively valued at $40,000,000 (4). Foresters and entomologists believe that these epidemics resulted principally from the severe drought conditions that prevailed generally over the period. It has been found that during periods of moisture deficiency, competition between trees is greatly increased and losses mount rapidly. It is believed that the dense stagnating stands of reproduction that have developed on vast areas have aggravated the beetle losses tremendously as a result of their competition with the larger trees for the limited soil moisture.

4. Because of these ecological changes, which are continuing to take place, the fire hazard has increased tremendously. Fires, when they do occur, are exceedingly hot and destructive and are turning extensive areas of forest into brush fields.

Conclusion

It is obvious that the present policy of attempting complete protection of ponderosa-pine stands from fire raises several very important problems. How, for instance, will the composition of other reproduction be controlled? If ponderosa pine is desired on vast areas how, unless fire is employed, can other species such as white fir be prevented from monopolizing the ground? On the other hand, if it is decided to permit such species as white fir to come in under mature ponderosa pine, how much of the public's money are foresters justified in spending in trying to keep fire out? Even with unlimited funds, personnel, and equipment, can they give reasonable assurance that they can continue to keep such extremely hazardous stands from burning up? If they feel reasonably sure of this, can they then give assurance that the timber products of such stands will be more valuable than those that might otherwise be derived from ponderosa pine and will in addition justify the high protection costs?

How will the density of the reproduction and pole stands on other vast areas be controlled unless fire is employed as a thinning agent? Except in stands on unusually good sites and favored locations, artificial thinning operations are so expensive as to be entirely unfeasible. The present stagnated stands have obviously resulted from total fire exclusion. It is well to remember that "rain-forest" conditions do not prevail in the ponderosa-pine reg ion and that root competition for the available soil moisture is probably, therefore, a more immediately critical factor than competition for light. The ponderosa-pin region is not capable of sustaining a dense forest like the Douglas-fir and redwood regions.

How will the terrific pine-beetle epidemics that are continuing to ravage large portions of the ponderosa-pine region be stopped? Present methods of artificial control, even the promising method recently developed of removing the most susceptible trees in light selection cuttings, are but expedients of temporary benefit. The probable aggravation of these epidemics, caused by the reduced vigor of the trees due to excessive root competition for limited soil moisture, has already been mentioned.

It is the writer's opinion that investigative work to solve these problems should be undertaken immediately. He believes that unless the stagnating stands of reproduction can be thinned and the terrific destruction by pine beetles can be checked the success of sustained-yield forest management in the ponderosa-pine region is seriously threatened. The present deplorable and increasingly critical conditions in vast areas o f the region are proof that foresters have not solved the silvicultural problems of ponderosa pine, and to continue present policies will further aggravate an already serious situation. These conclusions are the result of some seventeen years of close association with the ponderosa-pine forests, during which time ample opportunity has been afforded to observe the developments discussed. How to correct this increasingly serious condition constitutes a growing challenge to the professional forester and is a job worthy of the .best minds in forest research.

Literature Cited

1. Jepson, Willis Linn. 1923. The trees of California. Univ. of Calif. Press.

2. Keen, F. P. 1937. Climatic cycles in eastern Oregon as indicated by tree rings. Monthly Weather Rev. 65:175–188.

3. ———. 1940. Longevity of ponderosa pine. Jour. Forestry 38:597–598.

4. ———. 1940. Western Forestry and Conservation Association. Forestry Policy Conference, p. 97.

5. Meyer, Walter H. 1934. Growth in selectively cut ponderosa-pine forests of the Pacific Northwest. U.S.D.A. Tecb. Bul. 407.

6. Show, S. B. and E. I. Kotok. 1924. The role of fire in the California pine forests. U.S.D.A. Dept. Bul. 1294..

Indian Fires as an Ecological Influence in the Northern Rockies

1982. *Journal of Forestry*. Vol 80. No. 10. 647–651

Stephen W. Barrett and Stephen F. Arno

Stephen W. Barrett is research cooperator, Systems for Environmental Management, Box 3776, Missoula, Montana 59806. Stephen F. Arno is forest ecologist at the Northern Forest Fire Laboratory, Intermountain Forest and Range Experiment Station, USDA Forest Service, Missoula.

Abstract: The importance of fire as an ecological disturbance in the Northern Rockies is well accepted. Lightning is generally thought to have been the main source of ignition prior to settlement by Europeans. But writings of explorers and pioneers mention deliberate burning by Indians frequently enough to warrant an investigation of its importance. Interviews with descendants of Native Americans and of pioneer settlers in western Montana suggest that Indian burning was widespread, had many purposes, but was generally unsystematic. Fire chronologies based upon scars on old-growth trees indicate that fire intervals within similar forest types were shortest near Indian-use zones. Comparisons of presettlement fire intervals with those calculated from modern lightning-fire records suggest that Indian-caused fires substantially augmented lightning fires over large areas. As dependence on lightning fires alone may not create or perpetuate certain desirable plant communities or stand conditions, prescribed burning may be needed.

Knowledge of fire history is useful as a basis for developing ecologically sound fire management. Fire history has now been reasonably well documented for (Arno 1980), but one nagging question remains: What was the importance of Indian-set fires prior to settlement by European-American in about 1860? If lightning was chiefly responsible for presettlement fire regimes, it could be utilized as the primary ignition source when managing certain wilderness and natural area ecosystems.

But what if Indian-caused fires were an important influence on forest succession for centuries or even millennia? Then management for presettlement vegetation would require knowing to what extent those fires contributed to the development of various forest communities. Lightning ignitions alone might not be sufficient for reestablishing and perpetuating presettlement vegetation patterns. Or, if the policy is to utilize only lightning rims, knowing the effect of Indian fires would be vital in predicting forest succession. Finally, knowledge of the scope and effects of Indian burning might be useful in silvicultural manipulation of some commercial forest types.

That Indian-caused fires were important in the ecology of some Northern Rocky Mountain forests is suggested by:

1. General anthropological findings on widespread Indian use of fire in North America (Stewart 1956);
2. Specific examples of Indian use of fire in other western coniferous forests (*fig. 1*);
3. Inexplicably large concentrations of airborne charcoal occurring only in the last 2,000 years of sediments in a western Montana bog (Mehringer et al. 1977);
4. High frequency of fires as manifested on fire-scarred trees in Indian-inhabited areas (Arno 1976); and
5. Recurring mention in early journals and by pioneer residents of instances where Indians had set fires. For example, Lewis and Clark reported four Indian fires during their 1805 travels between Gates of the Mountains (near Helena, Montana) and Carmen Creek (near Salmon, Idaho) (Thwaites 1959, vols. 2 and 3).

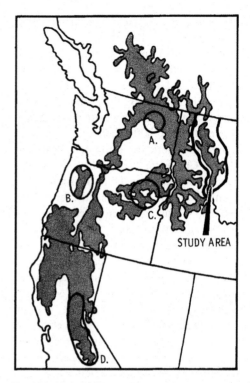

Figure 1. General distribution of the Pacific slope form of ponderosa pine, *Pinus ponderosa* var. *ponderosa* (after Little 1971); the map also shows the approximate distribution of mixed conifer-ponderosa pine types similar to those in western Montana. Lettered areas indicate locations where Indian burning has been implicated as an important ecological factor. (*A* Ross 1981, *B* Habeck 1961, *C* Shinn 1980, *D* Reynolds 1959.)

In 1979 a cooperative study was initiated between the USDA Forest Service's Intermountain Forest and Range Experiment Station and the University of Montana to determine the relationship of Indian-caused fires to the ecology of western Montana forests. The study resulted in an M.S. thesis (Barrett 1981) that serves as the basis for this article.

Montana West of the Continental Divide consists of grassland and forested valleys largely surrounded by heavily forested mountains. In much of this area the

forests at lower elevations are in the ponderosa pine (*Pinus ponderosa*), Douglas-fir (*Pseudotsuga menziesii*), or grand fir (*Abies grandis*) potential climax series (Pfister et al. 1977). At the time of European-American settlement, they were dominated by seral stands of ponderosa pine, Douglas-fir, western larch (*Larix occidentalis*), and lodgepole pine (*Pinus contorta* var. *latifolia*). Ponderosa pine, western larch, and large Douglas-fir are highly resistant to surface fires and often survived and prospered in stands that were burned at intervals of 5 to 30 years (Arno 1980).

Much Indian use occurred within and adjacent to these forest types. The Salish (Flathead and Pend d'Oreille tribes) and Kootenai Indians were the principal inhabitants of the area, and aboriginal occupation occurred for 6,000–10,000 years before European-American settlement (Malouf 1969, Choquette and Hostine 1980). Throughout that time the Indians numbered no more than a few thousand (Teit 1928), and had a hunter-gatherer economy. The advent of settlers about 1860, however, hampered the tribes' mobility, which was essential to their way of life. During the late 1800s the Salish and Kootenais were moved to the present reservation north of Missoula, Montana.

Evidence of Indian fire practices is sketchy. Barrett (1981) gathered information by interviewing older Indians and descendants of white settlers, searching journals and other written accounts, and analyzing fire-scarred trees.

Informants and Journals

We sought out the most knowledgeable descendants of the Salish and Kootenai and of pioneer settlers in the region and interviewed them by using the oral history methods developed by Baum (1974). Of 58 informants, 31 were Native Americans and 27 were of European descent. Responses concerning use of broadcast fire are summarized in *table 1*. The numerous "don't know" answers can be attributed to a dwindling knowledge of early practices, but the large proportion of "yes" responses and the lack of "no" responses suggest that broadcast fire was used by the Salish and Upper Kootenais. (The Lower Kootenais were a riverine-oriented culture, unique in this region because of their dependence on fish; they apparently made less use of broadcast fire than the other tribes.) Seven informants were able to give locations where burning by Indians occurred. In each case the informants' parents or an older acquaintance had witnessed the use of broadcast fire by Indians and had long ago related this information to the informant. These locations and others from journal accounts are shown in *figure 2*.

Testimony from informants and journal accounts indicates that Indian fires were most often set in low-elevation forests or grasslands and sometimes covered large areas. Fire history studies and early accounts and photographs in western Montana show that low-elevation stands in most areas were open and parklike (Leiberg 1899, Ayres 1901, Arno 1980, Gruell et al. 1982). These open stands evidently resulted from frequent low-intensity fires (underburns) that reduced fuels and understory vegetation. In the 1800s travelers often wrote that fires were widespread in the Northern Rockies and adjacent regions (e.g., Mullan 1861 and Phillips 1940). Similarly, a newspaper account (*Missoulian*, July 20, 1922, by Theodore Shoemaker of the USDA Forest Service, Missoula) of the early days in western Montana national forests stated that:

> Prior to 1897, and even later in

Table 1. Percent of informants according to their response to the question: "Did Salish or Kootenai practice deliberate burning?" (n = number of informants.)

Response	Salish (n = 38)	Lower Kootenai (n = 13)	Upper Kootenai n=7
Yes	44	15	71
No	0	54	0
Don't Know	56	31	29

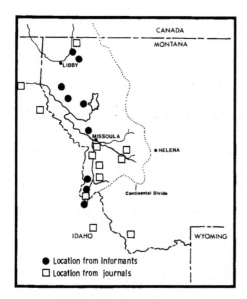

Figure 2. Locations of known Salish or Kootenai ignitions in western Montana and adjacent areas between approximately 1805 and 1920. Location references are listed in Barrett (1981, fig. 3).

1. Maintenance of open stands to facilitate travel, and clearing travel routes through dense timber;
2. Improvement of hunting by stimulating growth of desirable grasses and shrubs, to facilitate stalking, and to drive or surround game;
3. Enhancements of production of certain foods and medicine plants;
4. Improvement of horse grazing;
5. Clearing of camp site areas—reduced fire hazard and camouflage for enemies, and cleaning up refuse (this was the most systematic use of fire indicated by informants); and
6. Communication, by setting large fires.

Evidently the Indians of this region used fire informally to suit their immediate purposes rather than systematically as has been reported for tribes in California and Alberta (Reynolds 1959, Lewis 1973 and 1977).

Fire History Investigations
Fire frequency in heavy-use and remote stands

We reasoned that if Indian-caused fires covered much of the forest, sites in or near major travel and occupation zones would have been burned more often than similar sites in areas of infrequent use.

Ten pairs of old-growth stands were selected for fire history comparison. One member of each pair, the "heavy-use stand," was on slopes directly above a large intermountain valley, and the other," remote stand, "was on a similar site (in terms of aspect, elevation, and habitat type) several miles up a secondary drainage. Because of the steep, rocky, heavily forested terrain, only occasional hunting trips probably would have been made into the remote areas within recent centuries.

In each stand (200 to 600 acres) five to seven of the oldest fire-scarred trees were sectioned with a chain saw and a master

many sections, fires burned continuously from spring until fall without the slightest attempt being made to extinguish them.

Informant, journal, and anthropological information suggests that the Indians commonly used broadcast fire as a subsistence technology. Evidently fire was used informally and did not have religious significance. Explorers and settlers seem to have taken notice of Indian use of fire, because many considered such use unusual or hazardous.

Journals testify that unintentional and inadvertently set fires (including escaped campfires) were common Additionally, informants cited a number of masons for purposeful fires.

fire chronology was constructed (Arno and Sneck 1977). Mean fire intervals (MFIs), or the average intervals between fires, were calculated for each stand for three periods: pre-1860 (prior to settlement by Europeans); 1861–1910 (the "settlement" period); and 1911–1980. In the last period the USDA Forest Service and other organizations endeavored, with considerable success, to prevent and suppress fires. The objective of this research method was to examine changes in fire occurrence through these cultural periods.

Before 1860, MFIs were substantially shorter in 9 of 10 heavy-use stands than in their remote mates (*fig. 3*). Three of the heavy-use stand were in areas identified by informants as having been repeatedly burned by Indians. These stands had some of the shortest MFIs (seven to eight years). The one remote stand with a greater frequency of fire than its heavy-use mate may reflect poor sample stand selection; it was only three miles upstream from hot springs that may have been used by the Salish.

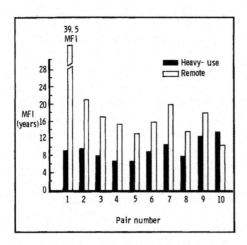

Figure 3. Mean fire intervals (MFIs) compared between 10 heavy-use stands and their remote mates during the presettlement period, prior to 1860. Generally the period of comparison begins in the early 1700s (modified from Barrett 1981).

Statistical testing (weighted t-test, Snedecor and Cochran 1967) indicated that three of the heavy-use stands had significantly (0.05 level) shorter MFIs than their remote mates. Also, the combined MFI for all heavy-use stands (9.1 years) was significantly shorter than that for the remote stands (18.2 years). The longer MFIs in remote stands resulted largely from a few especially long fire intervals, interspersed with various shorter intervals. Such temporal variation seems consistent with the chance occurrence of lightning fires. Heavy-use stands had short, more regular intervals suggesting a lightning-fire regime augmented considerably by Indians.

Fire intervals during 1861–1910 showed a pattern similar to that of the pre-1860 period, but differences between the heavy-use and remote stands are smaller. The overall MFI was still significantly shorter for heavy-use stands (11.5 years) than for remote stands (14.8. years). The inference is that Indian-caused fires were less frequent after 1860, although journals (e.g., Leiberg 1899) indicated that prospectors, settlers, and travelers caused many fires. In the fire suppression period, MFIs for all stands increased markedly, and differences between heavy-use and remote stands were not significant. Several stands contained no fire scars dating from that period; thus an overall MFI could not be calculated.

Frequencies of pre-1860 fires compared to modern lightning fires

Another approach was to compare presettlement fire intervals with those calculated from modern records of lightning fires. Lightning-fire occurrence was assumed to be comparable during both periods—(1) pre-1860, when lightning and Indians were the ignition sources, and (2) 1931–1980, when lightning was the major ignitor and detailed records of ignition sources were available.

Two areas were chosen for sampling, each being an east-facing slope adjacent to the Bitterroot Valley—the ancestral territory of the Salish Indians. These areas— Goat Mountain and Onehorse Ridge—are topographically isolated by rock-walled canyons. The steep rock terrain would have limited the probability of fires spreading from the south, west, or north, and surface winds are seldom from the east in this area. Only by selecting sites exposed to a minimum of fires from outside areas could the following comparisons be made. To determine pre-1860 MFIs, fire scars were sampled on 12 or more trees in each area and master fire chronologies were developed. Next, lightning-fire MFIs were estimated from 1931–1980 records of the Bitterroot National Forest. The records probably include every lightning ignition, for both areas are in full view of ranger stations and valley residents. The records allowed elimination of man-caused fires from consideration; moreover, both sites are undeveloped and have little human use.

The next step was to examine each lightning ignition to see if it had some potential to cause a spreading fire, had there been no suppression. The Individual Fire Reports (USDA Forest Service, Form 5100 series), along with weather, season, and rainfall before and after the ignition, were inspected to determine if much opportunity for fire development existed. The tendency was to be liberal, and most ignitions were accepted as having spreading-fire potential. At the Goat Mountain site (about 600 acres) the modern lightning-tire records indicated an MFI of 12.5 years, while MFI for 1700 to 1860 was 5.7 years. At the Onehorse Ridge site (about 1,000 acres), modern records indicated an MFI of 12.5 years, while MFI for 1631 to 1860 (data from Arno 1976) was 5.2 years.

Uncontrolled sources of error inherent in this approach preclude statistical testing. The pre-1860 MFIs undoubtedly include some spreading fires from outside these isolated sites. Conversely, some small fires no doubt missed the sample trees. Also, 50 to 60 percent of the modern lightning ignitions were projected to have had the potential to spread, and this may be a substantial overestimate. Data from 48 lightning ignitions (1973–1980) allowed to burn under prescription in nearby national forest wilderness showed that only 28 percent of those below 5,500 feet elevation developed into spreading tires (larger than one-quarter acre). Thus it appears that the sources of error tend to counterbalance each other and, overall, the results suggest that Indian ignitions contributed to the short fire intervals prior to 1860.

Fire Intervals and Acquisition of the Horse

It seemed plausible that Indian burning increased after the acquisition of horses about 1730, since horses increased the Indians' mobility (Roe 1955) and may have necessitated clearing forest understories for travel and for rejuvenation of grass. To test this hypothesis, we examined all fire chronologies that extended back at least 50 years before 1730 in heavy-use zones. These chronologies were obtained from individual trees or from several very old trees in a stand. We then compared relative frequencies in the pre- and post-1730 periods. MFIs were calculated for 19 trees growing in eight heavy-use stands and one remote stand. Master fire chronologies for entire stands would not be appropriate for this test, because the composite chronologies incorporate more individual-tree records during the later years, and thus would inherently yield shorter MFIs after 1730. The pre-1730 intervals averaged slightly longer than those from 1731 to 1860, but the difference was not significant at 0.05 or 0.10. Thus, it appears that Indian burning did not markedly increase after acquisition of the horse. Chronologies from small sites with the earliest fire records (see also Arno 1976) indicate that fire intervals remain similar back to the year 1500.

Some Implications

These investigations suggest that Indian ignitions substantially increased fire occurrence in lower elevation forests in and near the major valleys of western Montana. Fire-scar data show that most heavy-use areas lacked the long fire intervals that allow dead fuels to build up and understory trees to develop. Data from remote stands suggest that where lightning was the dominant ignition source occasional long fire intervals (30 to 70 years) resulted These long intervals would have allowed successional communities to become more diverse than in the heavy-use zones and would also have allowed fires to burn more intensely.

Sequences of photographs dating from 1898 to 1979 show successional change in a heavy-use zone forest in the Bitterroot Valley (Gruell et al. 1982). The pre-1900 MFI here was even years, and the resulting stands were dominated by large ponderosa pines with an open understory and herbaceous undergrowth. After 30 to 40 years of fire exclusion, willow (*Salix scoulerana*), bitterbrush (*Purshia tridentata*), other shrubs, and patches of tree regeneration increased. Today, after more than 80 years without fire, dense pole-sized understories of conifers (much of it relatively shade-tolerant Douglas-fir) have developed beneath the partially cut old-growth pine.

In many stands in the Douglas-fir and grand fir series in western Montana, long-term fire exclusion, with or without partial cutting, has now brought about dense overstocking and large, continuous buildups of fuel, particularly live-ladder fuels that could allow fires to crown and destroy the stand. The overstocking and shift in composition to more shade-tolerant species might also increase susceptibility to insects and diseases (Felhn 1979).

Managers of these forests might consider prescribed underburning soon after improvement cuttings of the overstory and slashing of excess or defective understory trees. Such underburns, using appropriate techniques and prescriptions (Fischer 1978, Davis et al. 1980), could reduce fuel and pathogen hazards, with little harm to pole and larger crop trees of ponderosa pine, western larch, and Douglas-fir. (Sometimes it may be useful to make an initial underburn prior to harvest or thinning to reduce accumulated fuels.) In addition to enhancing timber production, this approach has potential to stimulate growth of seral shrubs used as big game browse, and to result in esthetically appealing open stands (Gruell et al. 1982).

Implications for management of wilderness and other natural areas are that lightning fires may not be frequent enough to re-create presettlement conditions in the ponderosa pine, Douglas-fir, and grand fir series types Much of the terrain in wilderness is in zones remote from past heavy influence by Indians, but substantial portions, including major canyons and wilderness portals, are in the heavy-use zones. Also, many of today's natural areas and esthetic viewing areas are in past heavy-use zones. In both types of zones, reliance on lightning may be inadvisable because of the unprecedented long intervals since previous fires. Therefore, it may be necessary to set prescribed fires to achieve initial fuel reduction for returning some ecosystems to presettlement conditions. Such human-ignited prescribed rites in wilderness and natural areas may also be justifiable in terms of resuming an ancient approach of using fire to accomplish multiple objectives.

Literature Cited

Arno, S. F. 1976. The historical role of fire on the Bitterroot National Forest. USDA For. Serv. Res. Pap. INT-187, 29 p.

Arno, S. F. 1980. Forest fire history in the Northern Rockies. J. For 78:460–465.

Arno, S. F., and K. M. Sneck. 1977.A method for determining fire history in coniferous forests of the mountain west. USDA For. Serv. Gen. Tech Rep. INT-42, 28 p.

Ayres, H. B. 1901. Lewis and Clarke Forest Reserve, Montana. U.S. Geol Surv. 21st Annu. Rep., pt. V:27:80.

Barrett, S. W. 1981 Relationship of Indian-Caused Fires to the Ecology of Western Montana Forests. M.S. thesis, Univ. Mont., Missoula. 198 p.

Baum, W. K. 1974. Oral history for the local historical society. Am. Assoc. State and Local History, Nashville, Tenn.

Choquette, W. T., and C. Holstine. 1980. Cultural resource overview of the Bonneville Power Administration proposed transmission line from Libby Dam, Montana, to Rathdrum, Idaho. Wash. Architec. Res. Cent. Proj. Rep. 100, 63 p. Pullman.

Davis, K. M., B. D. Clayton, and W. C. Fischer. 1980. Fire ecology of Lolo National Forest habitat types. USDA For. Serv. Gen. Tech. Rep. INT-79, 77 p.

Fellin, D. G. 1979. A review of some relationships of harvesting, residue management, and fire to forest insects and disease. P. 335–414 in Proc. Syrup. on Environmental Consequences of Timber Harvesting. USDA For. Serv. Gen. Tech. Rep. INT-90, 526 p.

Fischer, W. C. 1978. Planning and evaluating prescribed fires—a standard procedure.USDA For. Serv. Gen. Tech. Rep. INT-43, 19 p.

Gruell, G. E., W. C. Schmidt, S. F. Arno, and W. J. Reich. 1982. Seventy years of vegetative change in a managed pondero pine forest in western Montana: implications for resource management. USDA For. Serv. Gen. Tech. Rep. INT-130, 42 p.

Habeck, J. R. 1961. The original vegetation of the mid-Willamette Valley, Oregon. Northwest Sci. 35:65–77.

Leiberg, J. B. 1899. Bitterroot Forest Reserve. U.S. Geol. Surv. 19th Annu. Rep., pt. V:253–282.

Lewis, H. T. 1973. Patterns of Indian burning in California: ecology and ethnohistory. Ballena Press., Anthropol. Pap. 1, 101p . Ramona, Calif.

Lewis, H. T. 1977. Maskuta: the ecology of Indian fires in northern Alberta. Western Can. J. Anthropol 7:15–52.

Little, E. L., JR. 1971. Atlas of United States Trees. Vol. 1. USDA For. Serv. Misc. Pub. 1146, 200 maps.

Malouf, C. I. 1969. The coniferous forests and their use through 9,000 years of prehistory. P. 271–290 in Proc. Coniferous Forests of the Northern Rocky Mountains. Cent. for Nat. Res., Univ. Mont., Missoula.

Meehringer, P. J., JR., S. F. Arno, and K. L. Peterson. 1977. Postglacial history of Lost Trail Pass Bog, Bitterroot Mountains, Montana. Arctica and Alpine Res. 9:345–368.

Mullan, J. 1861 Military road from Fort Benton to Fort Walla Walla. 36th Congr. 2nd Sess., House Exec. Doc. 44, 171 p. Wash. D.C.

Pfister, R. D., B. L. Kovalchik, S. F. Arno, and R. C. Presby. 1977. Forest habitat types of Montana. USDA For. Serv. Gen. Tech. Rep. INT-34,174 p.

Phillips, P. C. (ed.) 1940. W. A. Ferris: Life in the Rocky Mountains (Diary of the Wanderings of a Trapper in the Years 1831–1832). Old West Publ. Co., Denver, Colo. 365 p.

Reynolds, R. D. 1959. Effect of Natural Fires and Aboriginal Burning upon the Forests of the Central Sierra Nevada. Univ. Calif. Berkeley. M.A. thesis. 262 p.

Roe, F. G. 1955. The India and the Horse. Univ. Okla. Press, Norman. 434 p.

ROSS, J. A. 1981. Controlled burning: forest management in the aboriginal Columbia Plateau. Abs. 107 *in* Proc. 54th Annu. Meeting of the Northwest Sci. Assoc. Oregon State Univ., Corvallis.

Shinn, D. A. 1980. Historical perspectives on range burning in the Inland Pacific Northwest. J. Range Manage, 33:415–422.

Snedecor, G. W., and W. G. Cochran. 1967. Statistical methods. Ed. 6. Iowa State Univ. Press, Ames. 593 p.

Stewart, O.C. 1956. Fire as the first great force employed by man. P. 115–133 *in* Man's Role in Changing the Face of the Earth. W. L. Thomas, ed. Univ. Chicago Press.

Teit, J. A. 1928. The Salishan tribes of the western plateaus. Bur. Am. Ethnol. 45th Annu. Rep., P. 23–396. Wash. D.C.

Thwaites, R. G. (ed.) 1904. Original journals of the Lewis and Clark Expedition. 5 vols. Repr. 1959 by Antiquarian Press. N.Y.

The Politics of Wildfire: Lessons from Yellowstone

1998. *Journal of Forestry*. Vol 96. No 5. 4–9.

Pamela Lichtman
Reasearch Associate, Northern Rockies Conservation Cooperative

Land managers and ecologists generally agree that hte 1988 fires in the Greater Yellowstone Ecosystem were an ecologically important part of a natural disturbance pattern and that little could have been done to stop them. For policymakers, however, the fires were a major public relations failure. Land managers and ecologists need to understand how citizens' and politicians' view of wildfire as a crisis can undermine the stability of natural resource agencies, then find ways to build support for a natural fire.

Fire is a favorite subject of debate and investigation in the Greater Yellowstone Ecosystem, which comprises Yellowstone and Grand Teton National Parks plus seven adjacent national forests and other federal, state, and private lands in Wyoming, Montana, and Idaho (Clark and Minta 1994). Did fire policy fail in 1988? Did it irreparably damage our natural heritage and destroy commodities? Interest in these questions continues because of the dynamic debate over ecosystem management, but it surges whenever wildfires rage in the West.

The fires of 1988 may have been in evtable, judging from ecological evidence about fire return intervals in the Greater Yellowstone Ecosystem (Romme and Despain 1989). The political conflagration of 1988, however, could have been averted.

Developing a consistent and stable set of policies for land management into the next century will require broad public support for ecosystem-based initiatives and a realistic appreciation of the complexities of land management in a political context. It is not sufficient to consider just the technological d ecological dimensions of fire in natural landscapes., To build support for their programs, land managers and decision-makers need to understand how natural fire is perceived by citizens and then expand the idea of policy to include "packaging."

Antifire Sentiment
The 1988 Yellowstone fires officially began in late May with a lightning strike. The small blaze quickly died out, as many ignitions do under normal conditions. Others did not, however. Although as of mid-July

only 8,600 acres had burned, the National Park Service started suppressing all new natural fires. Whether the decision had any effect has been a source of debate, given that the largest fires and the majority of acres burned after mid-July, when severely dry conditions and gale-force winds provided ideal conditions for large wildfires. Cooler temperatures and early Septembers snows eventually brought the fire season to an end. All told, approximately 590,000 acres, or 45 percent of Yellowstone Park, burned in 1988, and roughly another 590,000 acres in surrounding national forests was somehow affected by the fires.

When the smoke cleared, many citizens and elected officials branded the government fires policy a failure and predicted grave results (Whipple 1993). Some people said the fires would "keep people away by the droves for a long time to come," and they lamented "the loss of wildlife and beauty... [and] economic potential" (Meyers 1988).

The fire policy, however, has been largely vindicated. Both scientific research and casual observations have shown that the actual effects of the 1988 fires differ greatly from what was expected. Wildlife a bounds, trees grow, and familiar historic buildings like Old Faithful Inn still stand. Studies of the effects of landscape-level disturbances in Yellowstone (Turner et al. 1994) have demonstrated that Yellowstone National Park was not "destroyed." Economic studies have shown that the fires did not impose excessive hardships on the region's human communities and may even have promoted growth in tourism by giving the park so much nationwide publicity (Snepenger et al. 1993).

The National Park Service and other land management agencies consider that vindication complete. The agencies withstood threats and invective from politicians, journalists, and armchair rangers, and the Park Service has been proven right about the effects on ecosystem components: grizzlies and geysers survived, elk and trees proliferated. Researchers' harrowing trip through the bottomlands of public opinion, followed by exoneration on scientific grounds, represents a major victory for the agencies. Scientists and land managers have mounted extensive campaigns—writing articles for general-interest media and scientific journals, producing videos, and convening conferences—to explain to citizens that the 1988 fires were a value-neutral ecosystem process. Yet in 1994, when wildfires burned in the western United States and provoked another massive fire-fighting effort, we saw once again the conflicting ways in which natural fire is depicted, interpreted, and understood in the United States.

Negative perceptions were common soon after the 1988 fires. Former Yellowstone Superintendent Bob Barbee and writer Paul Schullery (1989) observed that many descriptions of the postfire landscape relied on the metaphor of rebirth to suggest that the damage was not so bad after all. "Rebirth implies death," they noted, and in our culture, "death is evil. In the rhetoric of rebirth, Yellowstone has been killed by fires that must, by implication, have been evil, too." If the fires were evil, in the popular mind, then there should have been a policy to stop them. If the policy did not stop the fires, then it failed. By this logic, the Greater Yellowstone Ecosystem is now in a period of "rebirth" despite dire policies, not because of them. This message is one that people understand.

Examples of that outlook are numerous. In 1988 a *New York Times* editorial said there was "no reason to think this policy is in error" but offered tepid reassurance: "Yellowstone may take years to grow back exactly as it was, but even as soon as next spring the change may not seem so terrible (NYT 1988). This was hardly an endorsement of natural fire policy. In 1993 several area newspapers, including the *Jackson Hole News*, the *Billings Gazette*, and the *Casper Star-Tribune*, ran five-year retrospective articles on the 1988 fires. Writers expressed the sentiment that the fires were somebody's fault (Bellinghausen 1993; Thuermer 1993) and that Nature was being "reborn" in the aftermath of the fire.

Some sources were less equivocal about the meaning of the fires. *Country Living Magazine* (1994) heralded the planting of 482,008 trees "in our bid to reforest acreage destroyed in the Greater Yellowstone area during the fires of '88." Readers could pitch in by donating $15 to the reforestation fund (the actual planting sites were not specified). Other sources overtly indicted fire policy. A Jackson Hole businessman said,

> It's very shocking to see the degree of damage that was done. . . History has proved it was complete mismanagement. It was a tragedy and it could have been avoided. . . they've destroyed one of the greatest parks in North America (Thuermer 1993).

Even a former regional forester for the USDA Forest Service, defending clear-cutting policies on the Targhee National Forest, implied that 1988 represented a policy failure. John Mumma told the *Jackson Hole Guide*,

> I wasn't in favor of the Targhee's aggressive cutting. . . . But when I think about an alternative, I scratch my head. If the Forest Service had done nothing, the trees would have burned like parts of Yellowstone in 1988 (Welch 1993).

The implication is clear: incorrect policies doomed Yellowstone to burn, while proper management spared the Targhee. Different management strategies have different tradeoffs, though: hunting on the Targhee has been drastically curtailed because of lack of elk habitat, and the Targhee is now legally mandated to restore grizzly bear habitat (Welch 1994).

It is hard to judge just how average citizens now view the 1988 fires. Visitors to Yellowstone and other areas of the ecosystem probably get the message that the national park was not destroyed and that interesting things are happening on the ground as a result of the fires. But do they believe that these outcomes were in spite of fire policies, that the Greater Yellowstone Ecosystem is resilient in the face of "disasters," or that the fires simply were not

Forest regeneration was well established six years years after a 1985 burn in Grand Teton National Park; fire policy, however, remains subject to political whim.

as damaging as reported? If one of these reasons explains an average citizen's reasoning, then he or she may still see the 1988 fires as a policy failure.

What, then, did reviews of fire policy, research on fire impacts, and other responses to hte firestorm of criticism accomplish? The federal reviews of fire policy focused on fire management: forecasting, interagency coordination, and appropriate refinements of decision criteria (Fire Policy Management Review Team 1989; Clark and Minta 1994). The major conferences on fire policy since 1988 have focused on ecological effects. Social scientists and economist have studied the economic impacts in areas adjacent to Yellowstone, and other analysts have looked at media coverage. Each of these inquiries is important and adds to our knowledge about fires and their effects on landscapes and cultures. The vindication of the Greater Yellowstone fire policy, however, has occurred largely within academic and government communities. An understanding of the ecological function of wildfires still eludes the general public.

Political Feedback

The public's response to the 1988 fires was not to an ecological event but to the "construct" of an ecological event. In other words, people applied their customary frame of reference ot hte 1988 Greater Yellowstone fires and saw a crisis. Their construct—however understandable in a culture where people are weaned on *Bambi* and Smokey Bear, and no matter how unfounded it may be scientifically—has had real consequences for fire policy. It has also affected natural resource policy in general.

The most obvious consequence was the massive effort expended to fight the Yellowstone fires. Walt Thomascak, a Forest Service specialist (quoted in Wuerthner 1989, p. 50), offered this assessment of the costly suppression effort:

When the conditions get as bad as it was in the summer of 1988, you're wasting your money trying to suppress fires. But the public would never accept the fact that any federal agency merely stood around and watched a forest as it was consumed by flames, so the federal agencies have to put on a good show.

The firefighting effort was largely driven by public outcry from those who saw a policy failure. Although firefighters saved Old Faithful Inn, suppression in most cases was ineffective because of extreme fire behavior. For example, a human-lit backfire nearly destroyed Cooke City, Montana.

The agencies conducted a number of policy reviews after the 1988 fire season. Some resulted in minor adjustments, but none led to any substantive changes in fire policy. While the natural fire policy remained intact in principle, however, the impacts of political feedback may have altered the policy in practice. John Varley, chief of research at Yellowstone National Park (quoted in Wuerthner 1989, p. 43), summarized these impacts:

If we had to do it over again? Yes, I think we would do it differently. We would stomp on every single fire. Not because of the supposed "damage" done by fires, but because of the real damage done to people. Each of us has scar tissue that will last our lives. When I look at what people here had to endure in terms of public ridicule, and abuse from Congressmen, people in local communities, and the media people, who repeatedly attacked our integrity, intelligence, and professional ability—no, I don't think I or anyone else here would want to go through that again.

Knight (1989, p. 99) also saw political consequences: "If change in the policy does occur, the reason will probably be adverse public opinion rather than adverse ecological effects." Schullerty (1989, p. 53) expressed concern about the possible outcome of the reviews:

> A common thread in the current dialogue is the fear that while natural fire may be respected in principle, it will be eliminated in fact. It is easy to imagine that the political process may create a policy that in all respects is a model of approval—that expresses all the affirmative sentiments about the importance of allowing fire to play its natural role in the dynamic processes of the national parks—but that is so restrictive in its "circuit breakers" that in fact no fire of any useful size could ever occur.

Negative public perception also affected the way people perceived resource managers. If one believes that the 1988 fires were a policy failure, then one's trust in the land management agencies was betrayed. Although many who expressed a loss of trust in the agencies were perennial opponents of the federal bureaucracies, others were not. Neither group should be dismissed, since their opinion has ramifications for land management agencies' ability to function. During and immediately after the 1988 fires, there was a spate of denunciations of the agencies. One Montanan wrote, "I suspect there has been widespread negligence by federal, state, and county officials, with the citizens of Montana the victims" (Wheeler 1988). The commissioners of Park County, Wyoming, issued a resolution that "[t]he outright lies that were told to the public were unnecessary. It lost the public's confidence and damaged the agencies'

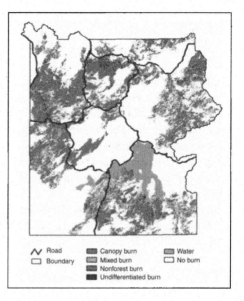

Figure 2. Ground conditions in Yellowstone National Park after the 1988 fire season.

credibility" (quoted in Oudin 1989). and the town council of Dubois, Wyoming, passed a similar resolution that chastised the Forest Service and the National Park Service for their "irresponsible attitude toward the public needs, and the futuer of this portion of Wyoming" (Dubois Frontier 1988). Charges of hubris, character flaws, collusion with special interests, and a host of other accusations about the land management agencies were broadcast to the public.

Although those views represent the extremes of antifire (and antiagency) sentiment, the rhetoric had real impact. The administrators of Yellowstone National Park felt hampered in their ability to move ahead with management initiatives. John Varley (quoted in Williams 1989, p. 84) commented on the damage that hte fire season had done to wolf recovery plans for Yellowstone:

> I think wolf reintroduction has been set back in a major way....The credibility of the agency has taken

such a nose-dive that our going out and saying, "Here's our program, trust us," isn't going to fly.

The outcry indicated that opposition to ecosystem management could quickly mobilize in the event of another fire or comparable management issue. Sanctions would not necessarily be formal, as in legislation or firings; retribution could come as denunciations in the press, budget cuts, or resistance to new programs.

Continual opposition and political sanctions may eventually condition the agencies to avoid certain activities and adopt "routinized rather than adaptive behavior" (Wilson 1989, p. 110). An aversion to risk may take root. If giving priority to ecosystem integrity becomes associated with negative consequences, then as a matter of survival, agency managers will treat those values as risky. Rather than engaging in the experimental, adaptive behavior that ecosystem management requires (Lee 1993; Clark and Minta 1994), manages may respond with rigid procedure more to avoid retribution than to conserve natural resources.

We already see a pervasive lack of faith in government to deliver goods and services—in areas like health care, education, and public safety. Brewer and deLeon (1983, p. 4) note that "profound social implications can occur as disillusionment with the government, its processes, and its general ability to serve the public good spreads." Such disillusionment is the cumulative effect of government's perceived shortcomings and citizens' unmet expectations—like the "failure" of the agencies to suppress the fires and "save" Yellowstone—and it undermines the long-term stability of the federal land-management agencies.

Improving Fire Policy

What options are available for fostering more accurate perceptions and "constructs"

of reality on the part of citizens, scientists, and policymakers? We must start by restoring public confidence in federal land management agencies, and then we need to develop processes that will allow for genuine public dialogue about the issues.

A realistic view of fire policy must acknowledge that clear rules for every conceivable eventuality are not feasible. An honest appraisal of how much control humans have over wildfires must be communicated to the public. and the ecological objectives of a natural fire policy should be persuasively presented to resource constituencies and policymakers.

There are a number of models for facilitating constructive involvement of diverse publics. Alternative dispute resolution is one possible model. The Forest Service's wilderness management program in the Bob Marshall Wilderness Complex is another model for participatory planning that includes input from citizens, land managers, and researchers. Political scientist James Fishkin has developed a deliberative opinion poll—a structured forum for introducing technical information into political deliberations of a small group of citizens. He finds that it "produces a voice of the people worth listening to" (quoted in Morin 1994).

Brunner (1994) recommends pilot projects that focus on tractable problems so that government can learn and prove its ability-to resolve issues and serve the public good. Designing a coordinated set of policies to deal with spatially discrete problems like the management of the Greater Yellowstone would be one such pilot project. The strategy would involve a number of technical policies for issues like wildland fire, grizzly recovery, and geothermal protection. Most important, it would allow informed. and competent citizens to help in policy formulation, decisionmaking, and implementation.

Selecting mechanisms for collective decisionmaking that best suit the d Greater Yellowstone Ecosystem is not an easy task. An important preliminary step is to identify active members of the policy community—such as elected officials, membership groups, stakeholders, and federal land management agencies—and collect some base line information about their beliefs, attitudes, and values.

Efforts should begin with small ale information-gathering projects around Yellowstone and throughout the policy community. Keeping deliberative processes small and decentralized maximizes citizens' chances to help develop policy. For example, it would be of little use to establish a citizens' panel for the whole Yellowstone ecosystem: the volume of site-specific issues would render the panel ineffective and unable to design policies for all areas, and the constituencies represented would be too numerous and diverse for a coherent voice to emerge. In any case, the practicalities of having citizens participate are daunting: there are no highways that directly connect one side of this huge region with the other.

Conclusions

Fire policy in the Greater Yellow one Ecosystem remains contentious. Fire is not "out there" or "on the ground" like grizzlies, elk, or trees. and when it does appear, people regard it as a crisis. Although the interrelated contingency plans that constitute current policy were scrutinized and refined in the wake of the 1988 fires, the possibility of error can never be eliminated from fire management.

Fire policy remains vulnerable to political repercussions. Just as a wild fire affects wildlife, timber, and tourism, fire policy has political impacts on large-carnivore policy, forestry, visitor management, and other issues central to Yellowstone; these impacts are inherent in the interrelated nature of ecosystem management.

No less than in 1988, the 1994 fire season provides a graphic illustration of how the political system responds to ecosystem processes. Although the ecological benefits of wildfires became clear it following the 1988 Yellowstone fires, the wildfires of 1994 were still fought with equal vigor. In addition, the 1994 fire season prompted further consideration of how natural fires ought to be handled. For example, the 104th Congress approved legislation in 1995 that dramatically increased salvage logging and even exempted such sales from environmental review processes because of the putative risk of wildfires (Kenworthy 1995). Clearly, a negative construction of natural fire is driving or being used to justify this and other natural fire policies despite compelling studies clarifying the important ecological effects the 1988 fires had on Yellowstone.

Land managers, ecologists, and sup porters of ecosystem conservation should understand that political feed a back has real consequences for land management policy. The term politics t itself embraces events and processes that involve human values and defy understanding, measurement, and explanation. Perhaps a fresh view of the things we pigeonhole as "politics"would lead to more effective participation in decisionmaking. Edelman (1988, p. 3) states that what makes political things political "is precisely that controversy over their meanings is not resolved. There is no politics respecting matters that evoke a consensus about the pertinent facts, their meanings, and the rational course of action." Our national parks and forests and natural fire do not evoke consensus, and thus they are political issues.

People who are involved in conservation of ecosystems cannot, then, dissociate themselves from politics. It is important to acknowledge and under stand how land management policies and ecosystem processes are interpreted and reinterpreted by the participants in our political system—citizens, their elected leaders, and the media. Recognizing that events like the 1988 Yellowstone fires get interpreted as a crisis is a critical step in designing more effective participation by citizens

The Yellowstone example reveals at the political repercussions on land managers and scientists are tangible and potentially damaging. Understanding how this process occurs and managing it will require not only effective communication with elected officials and public lands constituents nationwide but also mutual understanding and dialogue with people who are directly affected by fire, predation, and other ecosystem processes. Well-designed participatory structures incorporating the best available techniques from opinion polls, dispute resolution, and other specializations could increase effective participation by a region's residents.

Although it is important to be real tic and not expect to win near-unanimous support from residents for fire, wolves, grizzlies, and other controversies in the Greater Yellowstone Ecosystem, such efforts could at least produce better-informed disagreements. We will never eliminate "multiple realities and relative standards" (Edelman 1988, p. 111), but progress in ecosystem management cannot come until we make these realities less disparate.

Literature Cited

Barbee, R., and P. Schullery. 1989. Yellowstone: The smoke clears. *National Parks* March/April:18–19.

Bellinghausen, P. 1993. Yellowstone: After the fires. *Billings Gazette*, August 8:E1–2.

Brewer, G.D., and P. Deleon. 1983. *The foundations of policy analysis.* Homewood, IL: Dorsey Press.

Brunner, R.D. 1994. Myth and American politics. *Policy Sciences* 27:1–18.

Clark, T.W, and S.C. Minta. 1994. *Greater Yellowstone's future: Prospects for ecosystem management and conservation.* Moose, WY: Homestead Press.

Country Living. 1994. *Save our countryside.* August:16.

Dubois (WY) Frontier. 1988. *Resolution blames fire management.* July 28:1.

Edelman, M. 1988. *Constructing the political spectacle.* Chicago: University of Chicago Press.

Fire Policy Management Review Team. 1989. *Final report on fire management policy.* Washington, DC: Department of Agriculture and Department of the Interior.

Kenworthy, T. 1995. Babbitt finds relocation program has Hill's wolves growling at him. *Washington Post,* January 27:A23.

Knight, D. 1989. The Yellowstone fire controversy. In *The Greater Yellowstone Ecosystem: Redefining America's wilderness heritage*, eds. R.B. Keiter and M.S. Boyce, 87–104. New Haven, CT: Yale University Press.

Lee, K.N. 1993. Compass and gyroscope: Integrating science and politics for the environment. Washington, DC. Island Press.

Me [sic], C. 1988. Animal expert raps park fire policy. *Billings Gazette*, September 1:A10.

Morris, R. 1994. Thinking before they speak. *Washington Post National Weekly Edition*, May 16–22:37.

New York Times (NYT). 1988. *When to let the forests burn.* September 14:A13.

Oudin, D. 1989. Commissioners tear into 'outright lies' over wildfires. *Powell (WY) Tribune*, February 23:5.

Romme, W.H., and D.G. Despain. 1989. The long history of fire in the Greater Yellowstone Ecosystem. *Western Wildlands* 15(2):10–17.

Schullery, P. 1989. Yellowstone fires: A preliminary report. *Northwest Science* 63(1):44–54.

Snepenger, D.J., J. Johnson, and N. Friede. 1993. Tourism in Montana after the 1988 fires in Yellowstone National Park. In *The ecological implications of fire in Greater Yellowstone. Proceedings of the Second Biennial Superintendents' Conference on the Greater Yellowstone Ecosystem.* Yellowstone National Park, WY.

Thuermer, A.M. 1993. Fires of 1988 teach public about nature. *Jackson Hole News*, September 15:Al, Al9.

Turner, M.G., W.W. Hargrove, R.H. Gardner, and W.H. Romme. 1994. Effects of fire on landscape heterogeneity in Yellowstone National Park, Wyoming. *Journal of Vegetation Science* 5:731–42.

Welch, C. 1993. Grass roots action begins dialogue. *Jackson Hole Guide*, August 4:A8, Al0.

———. 1994. Targhee forest will cut logging. *Jackson Hole Guide*, February 2:Al.

Wheeler, R.S. 1988. Park's fire policy hidden in smoke. *Billings Gazette*, September 4:A9–10.

Whipple, D. 1993. Yellowstone recovers. *Casper Star Tribune*, October 3:El.

Williams, T. 1989. Special report: The incineration of lllowstone. *Audubon* January:38–85.

Wilson, J.Q. 1989. *Bureaucracy: What government agen cies do and why they do it.* New York: Basic Books.

Wuerthner, G. 1989. The flames of '88. *Wilderness* Summer:41–54.

Challenges to Educating the Next Generation of Wildland Fire Professionals in the United States

2009. *Journal of Forestry*. Vol 106. No. 107. 198–205.

Leda N. Kobziar, Monique E. Rocca, Christopher A. Dicus, Chad Hoffman, Neil Sugihara, Andrea E. Thode, J. Morgan Varner, and Penelope Morgan

Lela N. Kobziar (lkobziar@ufl.edu) is assistant professor, Fire Science and Forest Conservation, School of Forest Resources and Conservation, University of Florida. Monique E. Rocca (rocca@warnercnr.colostate.edu) is assistant professor, Department of Forest, Rangeland, and Watershed Stewardship, Colorado State University. Christopher A. Dicus (cdicus@calpoly.edu) is assistant professor, Wildland Fire and Fuels Management, Natural Resources and Management Department, California Polytechnic State University. Chad Hoffman (chadh@uidaho.edu) is instructor, Department of Forest Resources, University of Idaho. Neil Sugihara (nsugihara@fs.fed.us) is fire ecologist, US Forest Service, Region 5. Andrea E. Thode (Andi.Thode@nau.edu) is assistant professor, School of Forestry, Northern Arizona State University. J. Morgan Varner (jmvarner@humboldt.edu) is assistant professor, Wildland Fire Management, Humboldt State University. Penelope Morgan (pmorgan@uidaho.edu) is professor, Forest Resources, College of Natural Resources, University of Idaho. Leda N. Kobziar and Monique E. Rocca contributed equally to the preparation of this article.
The authors acknowledge the numerous students, fire managers, policymakers, and aspiring fire professionals who identified the imperative, contributed perspectives and feedback, and provided the substance for much of this evaluation of fire education in the United States. In particular, they thank the Association for Fire

Ecology and the International Association for Wildland Fire for providing professional venues for our special sessions on fire education during regional and international conferences in 2008. Special thanks go to the members of our panel discussions for their time and valuable insights, including Tom Zimmerman, Tom Nichols, Jesse Kreye, Mary Taber, Tobin Kelley, and the members of the Association for Fire Ecology Education Committee. The authors are grateful to Mark Koontz and two anonymous reviewers for their helpful comments, which improved the article. The views represented here are those of the Association for Fire Ecology Education Committee, and not necessarily those of their employing institutions.

Abstract: Over the last 20 years, the duties of US fire professionals have become more complex and risk laden because of fuel load accumulation, climate change, and the increasing wildland– urban interface. Incorporation of fire use and ecological principles into fire management policies has further expanded the range of expertise and knowledge required of fire professionals. The educational and training systems that produce these professionals, however, have been slow to organize an updated and coordinated approach to preparing future practitioners. Consequently, aspiring fire professionals face numerous challenges related to scheduling conflicts, limited higher education programs in fire science, lack of coordination between fire training and higher education entities, and the overall difficulty of obtaining education and training without sacrificing experience. Here, we address these and other challenges with potential solutions and outline the first steps toward their implementation. We organize the necessary aspects of professional fire preparation into a representative model: a fire professional development triangle comprised of education, training, and experience. For each of these aspects, we suggest changes that can be made by employers, educators, and nongovernmental organizations to provide a more streamlined mechanism for preparing the next generation of wildland fire professionals in the United States.

Keywords: fire management, fire ecology, education, training, firefighter

The past 20 years have been characterized by major development in fire science, management, and education in the United States (Stephens and Ruth 2005, Stephens and Sugihara 2006). Both wildland firefighting and fire management have shifted from supportive, ancillary roles (Greeley 1951) to positions of primary emphasis in many US land-management programs (Hiers et al. 2003). At the same time, in the US federal and state agencies that employ the majority of fire professionals, much of the workforce is at or nearing retirement age. Loss of the most experienced personnel is creating an increased demand for newly educated, trained, and experienced fire professionals, who are challenged by the growing complexity of fire management in the context of global environmental change, increasing wildland–urban interface (WUI), smoke impacts on human health, and other issues. Synchronous large fires have caused mass evacuations of residential areas, leading experts to ponder how fire's ecological imperative will be balanced with the protection of people and their property. The costs of fire management are high and increasing, making the effective education of future fire professionals critical. Recently, these developments have been accompanied by a proliferation of fire management vacancies and career opportunities, setting the stage for an evaluation of the educational systems that help prepare future fire professionals.

In this article, we, as the Education Committee of the Association for Fire Ecology (AFE), draw on over 65 years of collective experience as fire educators to assess the challenges that future fire professionals face. We have also conducted numerous informal interviews with fire management professionals from both USDA and US Department of the Interior (USDI) agencies; moderated panel discussions at International Association of Wildland Fire and AFE conferences; and solicited both written and oral feedback from aspiring and current fire professionals and students. Based on these discussions and our direct experiences, we offer our perspectives on improving existing and future preparatory systems designed to meet the evolving needs of fire professionals. The objectives of this article are to:

- Describe and assess the current context for fire professional development in the United States.
- Identify shortcomings with the current fire professional development paradigm.
- Explore potential solutions to the challenges we identify.
- Offer promising directions for innovation in preparing fire professionals.

We propose a new system that is predicated on cooperation between higher education providers and the various agencies and nongovernmental organizations (NGOs) engaged in training wildland fire professionals as they gain experience. The question, "How would we prepare the next generation of fire professionals if we were to start from scratch?" compels us to explore innovative solutions to the current challenges.

Fire management is in transition from an era dominated by fire suppression to one where fire use and suppression are equally viable resource management options (Stephens and Ruth 2005). Over the last few decades, fire has been increasingly incorporated into land-management programs as a component of ecosystem restoration and/ or maintenance, for fuels management, and for protection against the deleterious effects of wildfires on human and biological communities (Kilgore 1974, Parsons et al. 1986, van Wagtendonk 1991, Western Governor's Association 2001, Hiers et al. 2003). Fire management has become a designed combination of fire suppression and fire utilization, based on increased understanding of fire behavior and fire ecology (Sanderson 1974, USDA–USDI 1995, USDI 2001). Accordingly, the science on which sound fire management is based has grown in breadth and diversity, even as federal land managers are legally mandated to practice science-based management.

In light of the changing scope of fire management, the needs for professional staffing have rapidly expanded, outgrowing our current educational capacity and increasing demands for training. The types of education and training needed for future fire professionals have also evolved. In support of fire suppression, fire education has long focused primarily on fire as a physical process, on weather and fuel interactions, and on how to most effectively control fires (Gemmer 1979, 1980). With the increased recognition of fire's role in sustaining ecosystems and mitigating future wildfire risks, focus has expanded to include fire ecology and ways to use fire as one of many applied biodiversity conservation and ecosystem restoration tools. Furthermore, fire management is increasingly technology intensive, so that fire professionals must be adept at interpreting and applying the results of analyses based on remote sensing, geographic information systems (GIS), and models to support decisions. They must learn to evaluate which of many available tools is best suited for the task at hand.

Just as graduates need to be able to apply GIS, remote sensing, fire behavior models, and other technology (Zhao et al. 2005), they need to be adept at balancing social, economic, political, and ecological considerations (Sample et al. 1999). Such "broad and deep education" (Fisher 1996) could be well complemented with skills learned on-the-job through training and experience. For example, Gemmer (1980) proposed that university fire curricula be complemented by internships and by training courses through the National Wildfire Coordinating Group (NWCG). Others have called for educational changes to address the broad demands of forestry (Fisher 1996, Sample et al. 1999) and rangeland management (Kreuter 2001) professionals.

In this article, we focus on those professionals involved in wildland fire, including fire education, prevention, management, ecology, fuels management, and natural resources management. Fire professionals may specialize in fire behavior, effects, or management; ecosystem restoration and maintenance, fire suppression, and other tasks, (e.g., federal policy compliance). Fire professionals work in all five of the US federal land-management agencies, as well as in a vast network of fire-related positions with other federal, state, and local agencies; private contractors; and NGOs, including tribal lands management. Although individual job descriptions vary widely, future fire professionals must understand the multiple facets of fire's ecological role, be able to fore- cast and evaluate fire behavior and effects, and have direct experience with fire's impact in multiple ecosystems (Interagency Fire Program Management [IFPM] 2008). To stay current, this new generation of fire professionals must continuously

Education

Figure 1. The fire professional development triangle depends on integrating training, education, and experience to provide the background for achieving effective fire science and management.

incorporate new knowledge of fire ecology, fire behavior, and social sciences to tackle the multifaceted issues they will face.

To be effective, fire professionals therefore need training, experience, and education, all crucial parts of the fire professional development triangle (Figure 1). Working fire professionals need training to develop and maintain specific skills, knowledge, and competencies for operating equipment, managing personnel, administering complex fire management programs, and other job requirements. Training prepares the fire professional for standard fire use and research procedures, promotes safety awareness, and builds specific leadership and technical skills. Education couples an understanding of the behavior and ecology of fire with the ability to think and communicate creatively and critically, interpret complex information, and solve problems across multiple disciplines along various temporal and spatial scales. Experience continually expands and refines both education and training. To make sound decisions, fire professionals must reflect and draw on a breadth of experience with multiple fire events in diverse fire environments. Ideally, these elements will be part of life-long learning and integrated effectively. Their relative importance will vary depending on job responsibilities and stage of career.

Table 1. A sampling of universities and colleges with fire science programs in the United States (2008); included are those institutions that offer at least one certificate or degree in fire science and/or host an active chapter of the Student Association for Fire Ecology.

Institution (State)	Academic major	Degree seeking UG option/ concentration	Nondegree seeking UG certificate	No. fire-specific continuing/distance/ short courses[a]	No. semester fire-specific courses, UG	No. semester fire-specific courses, G
California Polytechnic State University (CA)	N	Y	N	3	8	1
Clark University (MA)	N	N	N	0	1	1
Colorado State University (CO)	N	Y	Y[b]	4[b]	4	2
Duke University (NC)	N	N	N	0	0	1
Humboldt State University (CA)	N	Y	Y	6	6	2
Louisiana State University (LA)	N	N	N	0	1	0
Mississippi State University (MS)	N	N	N	2	2	0
Northern Arizona University (AZ)	N	Y	Y	3	3	1
Ohio State University (OH)	N	N	N	0	1	0
Oklahoma State University (OK)	N	Y	N	0	3	2
Oregon State University (OR)	N	Y	N	1	3	2
Stephen F. Austin St. University (TX)	N	N	N	0	1	1
Texas Tech University (TX)	N	N	N	0	2	2
University of California–Berkeley (CA)	N	N	N	0	1	1
University of California–Davis (CA)	N	N	N	0	1	0
University of Oregon (OR)	N	N	N	0	1	1
University of Florida (FL)	N	N	N	1	1	1
University of Idaho (ID)	Y	Y	Y	8	6	1
University of Montana (MT)	N	N	N	0	1	1
University of Nevada–Reno (NV)	N	N	N	0	1	0
University of Washington (WA)	N	N	N	0	1	1
Utah State University (UT)	N	N	N	1	1	0

"Number of fire-specific courses" denotes courses with "fire" in the title. Table completed to the best of our knowledge as of February 2009.
[a] This column includes both special offerings and regular university courses that are offered in an alternative format (condensed, online, and so on) in effort to target midcareer professionals. Content may be identical to courses in the next two columns. This does not include NWCG courses offered through cooperation with the educational institutions.
[b] Technical Fire Management, offered through the Washington Institute, credit through Colorado State University during 1985–2008.
UG, undergraduate; G, = graduate; N, no; Y, yes.

Current Capacity of US Fire Education

Only a handful of the country's thou- sands of universities and 2-year and 4-year institutions provide substantial educational opportunities in wildland fire management and fire ecology. Fire management and ecology education has historically been concentrated in land-grant universities and technical community colleges, particularly those with forestry and range management programs, and is concentrated in the western United States where large fires are legend (Table 1). Furthermore, programs whose graduates are employed primarily by public agencies have had greater involvement in providing fire education.

Programs of study leading toward a fire-related BS degree range from standalone academic majors, options, and focus areas within related majors, to academic minors and certificates (Table 1). Even in academia, coursework emphases can differ between regions based on the cultural history of burning, management history, and the focus of the department within which

fire is taught. Traditionally, schools in the southern United States have focused more on prescribed fire use than western schools, which have emphasized fire behavior and science and other fire-related subjects. Within the 4-year schools with stand-alone majors and options (Table 1), all are located in land-grant schools (9 of 73 nontribal land-grant schools across the United States) or those with established natural resource education programs (6 of an estimated 80 schools with such programs). The linkage to natural resource- focused institutions, although geographically limiting, provides ancillary coursework and access to supporting faculty who, while not fire specialists per se, may have worked extensively on issues related to fire. Without a focus on natural resources, colleges and universities with environmental or biological science programs may be less likely to offer fire-related education.

Recently, increasing student interest has compelled academic institutions to expand the number and variety of academic courses and programs available to educate fire professionals. At the university level, at least five institutions have formalized new wildland fire options/concentrations/majors over the last 10 years (Table 1). Some have developed new courses and options in wildland fire sciences to capture the growing demographic of students with interests in wildland fire. Graduate fire science education at the MS and PhD levels has also grown, in part because of the relative stability of fire research funding associated with the USDA/USDI Joint Fire Science Program. As an indicator of increasing interest in fire sciences, the Student Association for Fire Ecology (SAFE), founded in 1998, has expanded to an internationally recognized entity with over 90 members and 13 official chapters at universities and community colleges across the country (B. Watson, pers. comm. SAFE, May 15, 2009).

In addition to traditional semester-based courses, a number of universities have developed short courses, online or distance-education courses, and other programs to help accommodate acting fire professionals who seek courses for academic credit. Several institutions have employed campus-based courses in an accelerated format that typically spans a few days to weeks rather than a typical 10-week quarter or 15-week semester. Additionally, these courses are often linked to extensive pre- or postcampus work that is facilitated by communication among students with the instructor via the Internet

There is not only a growing academic interest in fire science; recent updates to interagency fire management job descriptions and qualification standards have produced a new group of experienced fire professionals in need of fire-relevant university level education. Following the tragic firefighter fatalities of Colorado's South Canyon Fire of 1994, an interagency task group was assembled to investigate how training and education could better prepare fire management professionals

for the complexity of their professions. The resulting *Interagency Fire Program Management Qualification Standards and Guide* (IFPM 2008) defined 14 fire management positions with minimum qualification standards. Six of these 14 positions, primarily mid- to upper-level fire management positions (GS-09 and above), were classified into the federal government's GS- 0401 "General Natural Resources Management and Biological Sciences Series." Although the details and application of this change from a technical series to a professional series is still being refined, it adds a minimum educational requirement of either (a) an undergraduate degree in biological sciences, natural resources, or related fields, or (b) a combination of education and experience that includes at least 24 credit hours of coursework in related fields. These qualification standards apply to new employees, those desiring promotion, and individuals who held impacted positions before reclassification. Consequently, demand for academic courses and programs to help this group of fire managers meet the GS-0401 standards has been and is expected to remain high.

Academic programs involved in meeting the educational demands of current and potential GS-0401 professionals include those institutions listed in Table 1. Many aspiring fire professionals gain education, training, and experience in cooperative internship or trainee programs such as the Student Career Experience Program (SCEP) and Student Temporary Employment Program (STEP) (USOPM 2009). These programs provide students with work experience with an agency while they attend school. Although the STEP program is short-term, SCEP students may be noncompetitively converted to career, term, or career-conditional appointments if a position is available after graduation.

The Challenges

A historical deficiency of coordination and communication among universities and agencies has inadvertently resulted in barriers that prevent simultaneous access to all three legs of the professional development triangle. Traditionally, higher education institutions exclusively provided the educational component, while federal, state, and local agencies, as well as The Nature Conservancy and 2-year technical colleges, dominated training and experience opportunities. Traditional students find it difficult to attain training or extensive experience from agencies, and existing agency personnel often can not readily participate in higher education without significant time away from work. Discussions with fire professionals across the United States suggest that the integration of education, training, and experience in programs developing fire professionals is challenged by the lack of a common vision and a coordinated approach.

Academic calendars regularly overlap with seasonal employment for prescribed burning and wildfire suppression, making it difficult for students seeking experience and for fire practitioners seeking education (Figure 2). Agencies may be hesitant to hire traditional semester schedule-bound students for seasonal fire crews because of, in part, student's late arrival and early termination, jeopardizing crew cohesion and safety. This challenge may even be exacerbated by climate change, which is predicted to further extend the western US wildfire season (Westerling et al. 2006) into academic calendars. Even if academic calendars and fire seasons did not overlap, there appears to be an inherent discrepancy between educational goals and the requirements necessary for entry-level agency positions. Higher education is largely intended to prepare graduates for management positions, but graduates can not achieve higher ranks, or even obtain many permanent entry-level positions,

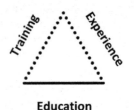

Education

Figure 2. Education acquired, but experience and/or training a challenge.

without agency-sponsored training and appropriate fire experience (Figure 2). Such experience provides the essential background from which critical fire management decisions can be made, and no amount of education can compensate for what experience bestows. However, training and experience often appear more important than education to students trying to secure both seasonal and permanent employment early in their careers.

Although many higher education providers have attempted to respond to this challenge by incorporating training and experience within the context of academic programs, the NWCG does not recognize university courses as meeting their specific training qualifications unless taught by an approved instructor. Few available instructors meet both university and NWCG requirements. Universities typically require lead instructors to hold a PhD, but acting professionals without a higher-level degree can give lectures and assist in teaching university courses. Even with such a team effort, it is unclear how to verify NWCG credit even if a university course explicitly covers NWCG-sanctioned material (using standard published guidelines, presentations, and exams). University courses often embellish the standard material with additional analysis, public speaking and writing assignments, and assessments of critical thinking. For example, in some universities, students plan, present, and execute a number of training- oriented tasks during prescribed burns. However, most often, students can not count these experiences toward meeting the qualifications required for post-entry-level positions in agencies, because an NWCG-trained and agency-employed representative must administer the training.

Once students have completed their undergraduate or graduate education in a fire-related subject, they must compete for available jobs. In many cases, the only available entry-level fire management positions are in firefighting, and lack educational requirements. Although critical thinking is an immediate benefit, in many cases the first promotion potential derived from education occurs in midlevel fire management positions, which may take more than 10 years of employment to reach. This scarcity of entry-level professional fire positions limits the potential of fire agencies to provide employment for students who complete higher education degrees.

Most fire organizations have traditionally hired young firefighters with little or no experience or university-level education. These individuals accumulate knowledge and expertise through experience, as they advance in their careers and gain training along the way (Figure 3). However, federal agencies have recently recognized the necessity of a broader educational base for many of their key fire management positions. The conversion of these positions to the GS-0401 Professional Series necessitates successful completion of a degree or 24 semester hours of higher education coursework. Many current employees holding these positions have obtained the required coursework via creative combinations of on-campus classes, online courses, and university short courses. Although the effort required to fulfill these education requirements is significant, the rewards equal the challenge; employees can maintain their current positions and gain the potential to advance into other professional series positions.

However, some midcareer professionals, although motivated toward self-improvement, may not pursue these educational opportunities because of a lack of agency support for both time and costs incurred, particularly if they are not currently in but aspire to GS-0401 positions. Regrettably, many senior-level federal employees have expressed frustration and demoralization, and even plan to retire early rather than meet the requirements of the GS-0401 series. Unless those who aspire to GS-0401 positions continue to find access to and support for educational opportunities, the vacuum in qualified employees will increase as employees who obtained the GS-0401 educational requirements retire in coming years (Figure 3).

In addition, fire professionals with extensive experience and training who are pursuing higher education for the first time or who are continuing in a new field of study often run into barriers. In particular, many individuals are either unprepared for the nature of university coursework or have not been exposed to the prerequisite biology, ecology, math, and communication skills needed to be successful in upper-division academic courses. Thus, students who already lack the support and/or time to return to school must also consider the possibility that they will need to take remedial coursework or spend extra time studying

Figure 3. Training and experience achieved, but education still a challenge.

prerequisite materials to succeed in upper-level classes.

Adding to this challenge is the paucity of educational programs tailored to current and future needs of the fire profession. Unfortunately, with ever-tightening budgets, lack of financial resources will likely frustrate efforts to start new programs or expand existing ones, even with faculty desire and a large student demand. Although the IFPM recognizes the value of higher education, there is no joint agency–academic consensus as to what constitutes sufficient academic standards for a fire professional. Without such an accreditation process, students are left without a clear pathway that would guarantee their preparation for professional fire positions and are thus less likely to pursue fire education. Unlike forestry and range science, which have academic accreditation standards dating to 1935 (administered by the Society of American Foresters and Society for Range Management, respectively), fire ecology and management are recently emerging fields that lack a single, overriding organization to develop such standards. In an attempt to start addressing this need, the AFE has recently proposed a fire ecologist/ fire professional certification program. The program is designed to foster a sound academic framework that encompasses the needs of diverse subdisciplines and geographic regions.

In addition to 4-year programs focused on wildland fire, a host of 2-year wildland fire programs at community colleges focus on technical education and training more than higher-order learning skills. Individuals who have received training through a 2-year technical degree program also face obstacles in obtaining experience due to the overlap between the academic calendar and seasonal employment opportunities (Figure 4). After degree acquisition, however, these individuals possess the required training to

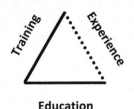

Education

Figure 4. Training and some education obtained, while gaining experience is a challenge.

compete for entry-level and seasonal employment, giving them an advantage over students completing nontechnical degrees. The 2-year degree students are thus (at least initially) better qualified to obtain fire employment where experience can be gained. Still, the 2 years devoted to education are at the expense of gaining experience in fire, and without a 4-year education degree students run the risk of not qualifying for certain professional series positions.

Additionally, challenges exist for technical degree students who wish to transfer into traditional 4-year programs to qualify for higher-level and GS-0401 positions. Students often lack the prerequisite courses required to continue studies in fire science or fire ecology. For example, a graduate from a 2-year wildland fire program may not have taken basic biology, chemistry, or physics classes required in a 4-year program. Typically, 2-year programs provide students with some basic educational skills, but emphasize fire-specific training activities. Technical degrees may not prepare students to transfer and complete 4-year degrees in only 2 additional years of study, potentially increasing the length of time over which education is traded for experience.

Overcoming the Challenges

The current structure of fire professional development programs makes it difficult for students to simultaneously achieve education, training, and on-the-ground experience. The current approach to gaining job qualifications is frustrating to aspiring and established fire professionals alike, who often find themselves with a lopsided "fire professional development triangle," with the weakest component impacting their ability to compete for jobs. We believe that this problem is so pervasive that it may soon limit the ability of our profession to respond innovatively and effectively to growing fire and fuels problems in the United States. New models for fire professional development are needed to integrate the three sides of the triangle, as well as restructure the professional development process. We describe our vision for a streamlined, integrated program, and recommend practical implementation steps to overcome the three challenges. Our aim is to ensure that wildland fire career building is more accessible, efficient, and effective over the long-term.

The Perfect Triangle: Our Vision for a Successful Professional Development System

The ideal system for preparing the next generation of fire professionals would integrate and/or provide in parallel education, training, and experience. Such a system would share characteristics with educational models used in other professions such as law, business, and medicine, where coursework is offered in conjunction with summer job experiences, training courses, and extensive internships. We suggest that a diversity of programs with different configurations ranging from 2-year technical programs through graduate programs be available to ensure that various career pathways are well paved.

Incorporating training and experience together with traditional coursework will likely necessitate a longer time commitment (i.e., an integrated program equivalent to a Bachelor's degree may take 5–6 years to complete). For such increased

lengths of programs to be palatable, a system-wide commitment to valuing the education, training, and experience obtained is essential. For example, graduates will qualify for positions commensurate with their integrated education and will have an advantage in competing against those who have not acquired the same degree of education. Therefore, the training, education, and experience components of these programs must be adequate to fulfill or exceed the IFPM-derived education and experience requirements of the target positions.

We propose that the first step toward resolving the challenges with the present systems of fire education is to foster open dialogue between the agencies that hire fire professionals, the developers, and instructors of NWCG training programs, and the higher education providers that represent degree programs. Improved communication will, undoubtedly, lead to innovative and mutually beneficial approaches to educating the next generation of fire professionals. Such coordination will have several important consequences: a mutual understanding and respect for what agency training programs and academic classes offer, with an appreciation of who is best equipped to provide each; a collaborative atmosphere revolving around a shared mission, as everyone works toward a common goal; and restructured educational systems that reduce redundancy and make education more accessible to both aspiring and midcareer fire professionals.

An important element of streamlining the professional development process includes the coordination of course content between universities and the NWCG. This coordination would result in increased access to NWCG training courses for aspiring fire professionals and increased access to university courses for midlevel fire professionals. By developing mechanisms by which agency instructors and university professors could co-teach selected courses, (such that students have the ability to receive both NWCG certificates and university credit), neither the students nor employers would have to pay twice (in time and/or money) for course content. Some universities already pair with agency instructors to make this possible, and student, agency, and academic response has been overwhelmingly positive. We emphasize, however, that we are not viewing NWCG courses and university courses as equivalent or substitutable. For credit in both academic and NWCG systems, students must be proficient and meet the goals of both entities.

Although we support streamlining the ways education and training courses are offered, we contend that agencies have a greater need and capacity to create and administer training courses, while universities are the more appropriate proprietors of academic tutelage. Where the content and goals of training and academic courses diverge, universities should take the lead in designing and teaching additional fire-related courses that have a true educational component (i.e., not pure training courses). Agencies should administer and teach the courses that involve specific skills needed to perform their jobs, such as safety and practical skills, and coordination between universities and agencies can ensure that students have opportunities to pursue both.

Ideally, a student in a university program who aspires to a fire professional position would have the opportunity to earn both the academic qualifications and the fire experience needed to qualify for a position on completion. We suggest that, simply through better coordination and collaboration between university faculty and agency personnel, many of the scheduling conflicts we have identified could be overcome, resulting in novel partnerships that nurture students and provide coordinated opportunities for learning.

New prospects for facilitating experience for university students might include designating some positions on seasonal crews specifically for students, offering trainee or intern positions for assisting with off-season management activities (e.g., planning and monitoring ongoing agency projects), or agencies inviting university students (courses, student organizations, or individual students) to participate in or observe prescribed burning and other management activities. Although many universities currently take advantage of the latter, it may be mutually beneficial to formalize these arrangements so that they are recognized as part of the future fire professional's accumulated experience and training. Alternately, students could take time off campus midway through their studies, after they have acquired a basic understanding of fire behavior and fire effects through their coursework, and spend a full season or longer working professionally in fire management.

If such traineeships were sanctioned as integral to the educational goals of the academic program, professors would adjust the academic calendar and curriculum to accommodate and reward such internship opportunities. Fire practitioners, in turn, would be formally mandated to mentor university students in the program. Participants would benefit from learning different aspects of fire management from their professors and from their typically more-experienced professional mentors.

For the special case of experienced fire professionals in need of academic credit, we propose that university coursework be considered an opportunity for growth. It should not be a duplication of training, but rather a mechanism by which professionals hone their critical thinking, analytical, writing, communications, and problem solving skills. We see many avenues to facilitate university education acquisition for experienced fire managers. Competitive leave or grant programs for undergraduate or graduate education would help agency employees take a "time out" from their careers to focus on academic classes. Alternative educational formats including distance learning courses and degrees, short course formats, and hybrids of these will make education more accessible and convenient for full-time fire professionals. These are already offered on some campuses (Table 1) and should be expanded.

Key Steps toward Implementation

We suggest the following developments and actions to hasten the improvement of the current system. The initial and most critical component is the establishment of an ongoing forum by which universities, agencies, and NGOs discuss common challenges, and the means to facilitate cooperation in achieving solutions to the challenges. The AFE Education Committee, Lessons Learned Center (an online discussion board resource), NWCG training committees, or a combined taskforce thereof would help organize this forum, which would guide and in some cases administer the following actions:

1. Establish shared standard expectations of future fire professionals between agencies and higher education providers. For example, the certification program developed by the AFE could serve as a springboard for discussions and potential strategies to achieve these standards. Such standards could guide the development and implementation of new BS programs, areas of emphasis, minors, or curricula in higher education to better meet the shared expectations of the desired future workforce.

2. Formalize agreements between higher education providers and agencies to bolster cooperation before, during, and after the education of future fire professionals. Develop and define viable career paths to ensure future fire professionals achieve experience, training, and education. Other opportunities for collaboration include agency input in course development, university access to training courses and experience opportunities, and mechanisms for career advising.

3. Enhance utilization and support for existing federal programs such as SCEP and STEP, designed to facilitate the transition from education to employment. This might also include enhanced opportunities for internships and traineeships that lead to permanent positions, financial incentives (e.g., fellowships awarded to graduates of fire programs) to encourage agencies to grant on-the-job experience to early career fire managers, and the establishment of a resume clearinghouse from which agencies can selectively recruit recent graduates of fire degree programs.

Conclusions

Universities and land-management agencies are partners in educating the fire professionals of the future. The professions of fire ecology, science, and management are greatly expanding in scope, breadth, and application. We are at a critical point in the transition from an emphasis on fire suppression to widespread fire management and use, resulting in the rapid expansion of professional workforce needs. Our current workforce is aging and we must provide an updated system that can respond to future needs and complexities. It is essential that the education, training, and experience programs that produce the future workforce are developed in a logical, thoughtful, and coordinated manner. Professionals in fire currently have the opportunity and the responsibility to provide a workable system for future wildland fire professionals. Given the multifarious challenges we face, we must work together toward meeting our common goals through a careful evaluation of the most effective distribution of responsibilities. We hope this article will inspire discussion and propel dialogue on this topic among the major organizations tasked with fire education and training. We recognize that there are numerous challenges, but, more importantly, see ample opportunities for improving our preparation of the future professional fire workforce.

Literature Cited

Fisher, R.F. 1996. Broader and deeper: The challenge of forestry education in the late 20th century. *J. For.* 94(3):4–8.

Gemmer, T.V. 1979. Forestry curricula today. *J. For.* 77:414–417.

Gemmer, T.V. 1980. Proposed curriculum for fire management professionals. J. For. 78:149–151.

Greeley, W.B. 1951. Forests and Men. Doubleday Publishing, Garden City, NY. 255 p.

Hiers, J.K., S.C. Laine, J.J. Bachant, J.H. Furman, W.W. Greene, and B. Compton. 2003. Simple spatial modeling tool for prioritizing prescribed burning activities at the landscape scale. *Conserv. Biol.* 17(6):1571–1578.

Interagency Fire Program Management (IFPM). 2008. *Interagency Fire Program Management qualification standards and guide.* Accessible online at www.nifc.gov/policies/ red_book. htm; last accessed May 2009.

Kilgore, B.M. 1974. Fire management in National Parks: An overview. P. 45–57 in *Proc. of the 14th Tall timbers fire ecology conf.* Fire and land management symposium, Oct. 8–10, 1974, Missoula, MT. US For. Serv. Tall Timbers Res. Stn., Tallahassee, FL. 675 p.

Kreuter, U. 2001. Preparing for the future of range science. Rangelands 23(5):24 –26.

Parsons, D.J., D.M. Graber, J.K. Agee, and J.W. Van Wagtendonk. 1986. Natural fire management in national parks. *Environ. Manag.* 10(1):21–24.

Sample, V.A., P.C. Ringgold, N.E. Block, J.W. Gitmier. 1999. Forestry education: Adapting to the changing demands on professionals. *J. For.* 97(9):4–10.

Sanderson, J.E. 1974. The role of fire suppression in fire management. P. 19–31 in *Proc. of the 14th Tall timbers fire ecology conf.*, Fire and land management symposium. Oct. 8–10, 1974, Missoula, MT. US For. Serv. Tall Timbers Res. Stn., Tallahassee FL. 675 p.

Stephens, S.L., and L.W. Ruth. 2005. Federal forest fire policy in the United States. *Ecol. Applic.* 15:532–542.

Stephens, S.L., and N.G. Sugihara. 2006. Fire management and policy since European settlement. P. 431–443 in *Fire in California's ecosystems*, Sugihara, N.G., J.W. van Wagtendonk, K.E. Shaffer, J.A. Fites-Kaufmann, and A.E. Thode (eds.). University of California Press, Berkeley, CA.

USDA–US Department of the Interior (USDI). 1995. *Federal wildland fire management: Policy and program review.* Final Rep., Dec. 18, 1995. 45 p.

US Department of the Interior (USDI). 2001. *Review and update of the 1995 federal wildland fire management policy.* Rep. to the Secretaries of the Interior, Agriculture, Energy, Defense and Commerce; the Administrator, Environmental Protection Agency; the Director Federal Emergency Management Agency; and the National Association of State Foresters, by an Interagency Federal Wildland Fire Policy Review Working Group. Boise, ID, National Interagency Fire Center. 78 p.

US Office of Personnel Management (USOPM). 2009. *Student educational employment program.* Available online at www. opm.gov/employ/students/intro.asp; last accessed May 2009.

Van Wagtendonk, J.W. 1991. The evolution of national park fire policy. *Fire Manag. Notes* 52: 10–15.

Westerling, A.L., H.G. Hidalgo, D.R. Cayan, and T.W. Swetnam. 2006. Warming and earlier spring increase western US forest wildfire activity. *Science* 313:940 –943.

Western Governors' Association (WGA). 2001. *A collaborative approach for reducing wildland fire risk to communities and the environment: 10-year comprehensive strategy.* Western Governors' Association. Available online at www.westgov.org/wga/initiatives/fire/final_ fire_rpt.pdf; last accessed May 25, 2003.

Zhao, G., G. Shao, K.M. Reynolds, M.C. Wimberly, T. Warner, J.W. Moser, K. Rennolls, S. Magnussen, M. Köhl, H. Anderson, G.A. Mendoza, L. Dai, A. Huth, L. Zhang, Liangjun, J. Brey, Y. Sun, R. Ye, B.A. Martin, and F. Li. 2005. Digital forestry: A white paper. *J. For.* 103(1):47–50.

Part 5

The Utilization of Fire

◼ Kelly Martin

The social and political construct of fire will continue to challenge manager's best intentions to allow fire's rightful place on the landscape. Our challenge is to protect the lives of firefighters and the public and the homes of people who have chosen these wildlands as places to live. Our children are inheriting a very different fire landscape than when I started my career over thirty years ago.

Yosemite National Park, where I work, is a small microcosm of a much larger narrative of why fire plays an essential function for healthy ecosystem services. Lightning fires are ubiquitous on the park's landscape every year. The science and research supports the use of fire as a natural and essential element of a healthy, functioning ecosystem. What is less obvious and much more complicated is the social, political, and economic impacts of fire within our local communities and throughout the United States. Four important pillars of fire management—ecological factors, social dynamics, political

inputs, and economic drivers—vary across the entire United States. Having strength in these four areas foreshadows the success or failure for managing both wildfires and prescribed fires.

The ecological niche of fire has the greatest amount of research and science to support managers when they view the tapestry of the landscape. During drought years, wildfires are allowed to burn in the higher elevations of Yosemite National Park more often than not. During normal years, many of these areas are too wet to burn, but during drought years a lightning strike finds a toe hold and thus begins our management for a long-term wildfire event. This yearly cycle of allowing some natural ignitions to burn throughout the summer season began under the academic and scientific leadership of Jan van Wagtendonk almost fifty years ago.

The social construct of fire begins with our own willingness as managers to shift our cultural norms from one of a suppression industry, to one that views fire as a potential protector, especially in the context of future climate change scenarios. As land managers, we cannot escape the escalating cost and size of contemporary wildfires. Fuel treatments and prescribed fire can and should be applied in and around our greatest values to tame the future oncoming wildfire. However, no longer should we accept the notion that we can thin and prescribe burn our way into the future as the sole manner of preparation. Wildfires are an inevitable truth. No less apparent is that no amount of fire suppression resources will be able to hold back the tide of what is yet to come.

Social acceptance for the various effects of an active fire is going to be required. As much as the public may dislike smoke it is an inevitable truth of living in and visiting fire-prone landscapes. We have stored an enormous amount of smoke on our landscapes due in large part because of our social acceptance of fire suppression to placate the innate human fear of fire. A societal shift from one of wildfire suppression to one of wildland fire management is desperately needed so that managers can implement sound fire ecology principles.

The political rhetoric of fire suppression versus using wildfire to meet land management objectives will provide a continuous flow of material for social media. So long as our society is attracted to the misfortune of others, those largest, most destructive, most expensive wildfires will remain as the marker for which all other fires will be judged. We desperately need the political will to support the wise use of fire in well thought-out, fire-adapted ecosystems. We need our political leaders to educate themselves or allow themselves to be educated to the true nuance that exists on the landscape. The truth is that some fires will always need to be suppressed. However, we are seeing more and more that there is more to the story, and that the utilization of wildfire and the application of prescribed fire to benefit

landscape resources is just as important as suppression, if not more so.

The final pillar, that of economic drivers as they are influenced by wildfire, cannot be overstated in its importance. This may take the form of a river guide needing clean water, or the general store owner depending upon campers to buy last-minute marshmallows and firewood, or the backpacker looking for solitude in a green landscape. These examples and more are real considerations to be made when determining what we wish our vegetated landscapes to afford us.

Landscapes that are burned severely, in ways not seen in the ecological record, often produce effects that are not supportive of positive economic factors. Watersheds that are burned to bare mineral soil, timber stands that suffer excessive mortality, soils that have their seed banks destroyed: these scenarios carry with them consequences for economics whether it be for recreation, for wood products, or for indirect benefits such as that general store owner who sells marshmallows to folks passing through, much less that pleasing and intact tree line to watch a sunset fall into.

The best way to reduce the impact to these economic drivers is of course to maintain the function of the landscape. We can debate endlessly about what one landscape should look like, especially in the era of climate change, however most would agree that burns that mimic the characteristic—call it historic if you prefer—are the burns that will maintain the relevant economics.

These four principal factors are woven throughout the five papers that have been selected for the final topic area of this book on the utilization of fire. Raymond Conarro's simple, straightforward, but powerful paper where "prescribed fire" is introduced both as a term and a concept speaks primarily to the economic importance of the southern pine lands, however he immediately also discusses how there should be a "realization that interests other than trees exist on a high percentage of southern forest lands, and that those interests deserve recognition..." In this paper also he writes about the social factors that persisted in the south; the often-told story about the "crusader exclusionist" and the "wanton woods burner," and how to blend that social conflict into a usable middle ground. In this case: the artful application of fire.

The other interests Conarro inferred include what the second author, Herbert Stoddard, concerns himself with directly in his paper about the management of upland bird habitat (sporting birds specifically). Stoddard was famously not a forester, but rather a wildlife biologist. The problem, of course, was that the habitat for his birds was the forest, and the forest was managed by foresters who did not want to burn the woods. Stoddard saw early evidence that regular burning of the woods did indeed produce positive results for habitat and other ecological factors. The result, as he saw it, was that healthy forests meant a place where landowners could hunt (or profit from hunting), allow the continuation of a recreational pastime, and seemingly keep the forest healthy too.

Chapman's paper specifically calls out the federal and state land managers for their recalcitrance toward the application of prescribed fire on the landscape. By 1947, when this paper was published in the *Journal*, he could be more forward in his highlighting of the obvious ramifications that would occur should the U.S. Forest Service continue to bury its head in the sand, as he likens them to doing similar to the British Forest Service in India. He makes a brilliant analogy to previous experience—and failure— to recognize the role of fire for what it is. Strengthening the argument is his tie into the four pillars we identified here.

Fast forward almost thirty years and we arrive at both Robert Cooper's and Bruce Kilgore's papers from 1975. Both papers reflect that the tide in the forestry profession has shifted in favor of using prescribed fire for a desired result, however these are still the early years of widespread direct implementation. It is fascinating to see how long it has taken for acceptance to occur and for the practitioners to have the latitude to experiment on their own units.

Cooper's paper bears note because it was written by someone from the Southern Research Station, long the nexus of prescribed fire research, discussion, and argument. The author writes in straightforward language about the positives and negatives of prescribed fire. He also advocates for taking the lessons learned from the South and seeing how they can be applied elsewhere in the country. This is a pivotal time for the western fire scene, with few people outside of Harold Biswell really "picking up the torch" and looking to implement prescribed fire somewhere other than the southern pine forests.

If Biswell embodied an evolution of Chapman, then the final author embodies an evolution of Biswell. Fortunately the work of Chapman and Biswell, which fought against the status quo, inspired many others to follow in their iconoclastic footsteps. The last paper describes the fruits of these efforts in a seminal story of ecological restoration in the Sequoia groves.

Authors Bruce Kilgore and Rodney Sando were disciples and early implementers of active ecological restoration work, and they were not alone. I would be remiss not to mention the importance of other researchers who also rose up in defiance of the status quo for what they believed was best for the landscape. Most influential in Yosemite is Jan van Wagtendonk, whose career and impact is equal in every way to researchers such as Chapman and Biswell, and who has guided the hand of many fire managers who have grown under his tutelage.

The Sequoias of the Sierra Nevada were one of the first places to see active restoration activities in the west outside of the Ponderosa pine belt, and prescribed fire was an intimate and necessary component. Fortunately, the Sequoias, because of their majesty, garnered enough political and social support that saving them was a priority for everyone. The work of Biswell, Kilgore, and van Wagtendonk helped become the catalyst for prescribed fire to be viewed as a tool throughout the west, in many different ecosystems.

The work of understanding and implementing prescribed fire, or more broadly the utilization of fire, is not complete. Many contemporary managers are seeing the very real need to incorporate not only prescribed fire, but natural fire, and allowing it to perform its work on the landscape. Just as the implementation of prescribed fire had its fits and starts (we're *still* learning lessons), so will this next step. We will without question be tasked with gathering lessons for how to use wildfire on our landscapes correctly. The challenge for us is to accept a certain level of risk today in the wise use and application of fire in an effort to thwart an even risker and more dangerous fire environment in the future. Every last one of should believe we share a vested interest in managing fire for successful outcomes, and we should strive to approach this belief with a humble understanding that fire management is just as much about science as it is art.

Use of Controlled Fire in Southeastern Upland Game Management

1935. *Journal of Forestry.* Vol 33. No. 3. 346–351.

 Herbert L. Stoddard
Director, Cooperative Quail Study Association

There are, to my knowledge, approximately a million acres in the states of South Carolina, Georgia, and Florida alone, with a large but undetermined aggregate acreage in the remaining states of the southeastern group, owned primarily for the shooting of quail and other upland game. There is an even greater acreage where the renting of shooting rights bring landowners from ten to fifteen cents per acre annually. Both the owned and leased acreage mentioned, is in game preserves largely of northern and eastern sportsmen, who spend large sums in the communities where their hunting lands are located. Incidentally these properties usually carry a much heavier stock of game than surrounding unprotected lands, hence provide an overflow of game for public shooting.

In addition to the acreage in recognized preserves, quail shooting on both public and other privately owned lands furnishes one of the most popular sports of the South.

Hence it is evident that quail are an important economic asset, especially from the recreational standpoint.

In bygone days an abundance of these fine game birds was largely taken for granted in the region, for they thrived in connection with the favorable agriculture and land use of the country, while not as extensively hunted as at present. During recent years, however, some attention has been given to increasing quail and other upland game through game management practices, for it is becoming recognized that only by such means can a supply be maintained in the face of increasing demand.

Obviously, the success of game management will in the long run, be largely in proportion to our knowledge of the general life history, feeding habits, diseases, parasites, enemies, cover requirements and preferences, and so forth, of the game being managed. Need for dependable information on quail was responsible for the Cooperative

Quail Investigation of 1924–1930, which was financed by quail sportsmen, and conducted by the United States Biological Survey. Desire for additional information and *how to apply it to the land*, has been responsible for the Cooperative Quail Study Association, which in 1931 resumed studies and experiments where the earlier investigation left off. The present work is being conducted by the speaker under private auspices, though continuing close cooperation with various branches of the United States Department of Agriculture, and other public agencies.

Studies of the effect of periodic burning of ground cover on quail, and incidentally other wildlife, has from the first taken a prominent part in our programs. To suggest as a result of these studies, that fire has a legitimate use in the improvement or maintenance of game range, may come as a shock to some who believe fire a curse under all circumstances. The importance of this agency in southeastern game management is such, however, that a brief discussion seems warranted here, for it is generally misunderstood, and much misinformation has been circulated in respect to it. We would like it clearly understood that we are recommending controlled use of fire mainly on quail and wild turkey ground, where an abundance of these game birds is considered of first importance by owners of the land. We are also confining our discussion to the open pineland type of forest characteristic of much of the south Atlantic and Gulf Coastal Plain, though use of fire in upland game management may be as essential elsewhere. and it should be obvious in this connection, that the subject of the relation of fire to wildlife is so complex that we can do little more than touch upon it in the time at our disposal.

We are perfectly willing to agree with advocates of fire exclusion that wrongly used fire is capable of damage to quail and other game interests in the region. Nothing for instance, can be more destructive to ground nesting game birds than summer fires which destroy nests and young, together with growing food supply and cover, and all conservationists should combine against them. We do, however, consider that carefully controlled fire, used at the *proper season, under proper weather conditions,* for the definite purpose of regulating cover and increasing food supply of the game birds, is a necessary tool over much southeastern game territory, and an essential feature of quail management in the region.

I may say at this point, that fire might conceivably be largely dispensed with even in quail management, were it possible economically or physically, to periodically plow or harrow the bulk of nonagricultural ground where quail are desired. In most cases this is obviously impossible where the terrain is covered with forest, and a high percentage of our quail ground consists of open pine woodlands.

Our studies have made clear several reasons why quail especially, profit from properly used fire, some of the most important of which will be briefly presented. In the first place, these birds are comparatively weak scratchers, so dense tangles of wire grass or broom sedge, or accumulations of pine needles and other vegetable debris, exclude them from their food supply as completely as would an equal depth of snow. Such accumulations also, compete with and smother by mulching, a vitally important source of their winter and spring food supply, especially that furnished by native perennial legumes, and certain other herbaceous vegetations.

Quail also require for roosting, nesting and feeding range, ground cover that is open below, though furnishing some protection from winged enemies above. Other adverse effects of over density of ground cover may be of equal importance. For instance, cotton rats thrive in cover that is denser than that needed for quail, and where rodents abound, hawks, owls, skunks, foxes, house cats and wild cats, and several species of snakes may appear to feed upon them, and incidentally prey to a certain extent upon the quail. As far as we have learned regulation of ground cover density for our purposes can only be accomplished by the plow, or by periodic use of fire, for grazing with live stock is generally adverse to quail. For that part of the quail's food supply furnished by ragweed and other weeds and grasses which normally accompany cultivation, plowing at intervals is usually preferable to fire. For certain other feeds, especially Partridge Pea (*Chamaecrista sp.*), which is of outstanding importance to quail, either plowing or fire will serve, though fire is generally preferable, as well as more economical. For proper maintenance of ground cover, especially where grasses and litter have a tendency to rapidly accumulate, fire is often superior to plowing or harrowing, though an occasional plowing is usually advantageous if terrain and budget permit.

The *season* of burning is of great importance to some game food plants, though weather rather than the calendar should govern the time of burning for them. Broom sedge or grassy areas containing Japan clover, an annual lespedeza of great importance to quail, must be burned over early, *before germination takes place,* else eradication results. If burned early enough, this clover is greatly benefited by the removal of dead vegetations and litter, which in the absence of grazing, acts as a mulch and ultimately smothers it out. Selection of the season of burning for partridge pea is equally important. If burned as early as January the fire-scarified seed may germinate prematurely near the warm surface of the fire-blackened ground, and the resulting young plants be killed wholesale by March freezes. If burning is carried on too late,

however, such plants as may have started growth in the "rough" will be killed by fire. Both of these important food plants are annuals. Fortunately the native perennial legumes, which include the very important beggerweeds, lespedezas, or bush clovers, and many others, sprout vigorously and seed prolifically even if pruned back by fire as late as May, so burning for them may be carried on at any time from midwinter to late April.

It is important, in general, to conduct necessary burning as late in the season as is possible without injury to the nesting of the game birds, or the plants which produce their food supply, so that the former will have the protection of new cover as soon as possible after removal of the old.

Burning with the right *frequency* is as important as burning at the proper season, and this also depends upon such factors as luxuriance of vegetations as influenced by growing seasons or type and fertility of soils, as well as on *kinds* of foods being worked for. For example, where fruits from dwarf varieties of huckleberry, blueberry, and certain other shrubs, or "mast" from such ground vegetations as dwarf chinquepins and runner oaks are desired for the game, fire should not be used annually, unless indeed clumps of these producing plants are given special protection by fire lines, when surrounding ground cover is burned. This because such shrubs do not bear well the year of a burn, even though an occasional pruning back by fire is highly beneficial. Thus for wild turkeys, which draw heavily on this group for food, burning where such food plants are important may be desirable at rather long intervals, as every third or fourth year. On the other hand, native perennial legumes, which are of special importance to quail, bear best, and the seeds are more available to the birds when burning is more frequent. While observation has shown that doves and many other seed eating birds are greatly attracted by burns, the situation in this case has not been studied sufficiently for discussion here.

Burning is as necessary to quail hunting, and to high grade dog work on quail, as it is to the quail themselves. For pleasurable hunting, brushy undergrowth in the pinelands must be kept down by fire of the required frequency, if expensive hand work with brush hooks and axes is to be avoided. Dog work may also be poor where accumulations of litter are deep, which is a practical consideration in the management of hunting lands.

Granting that controlled fire may be the only practical agency for purposes enumerated, the question may well be asked "How can necessary burning be conducted for quail and other game in such a manner as not to unduly jeopardize such interests as forestry, where the latter is also of importance? Now as a mere game manager, I have some hesitancy in discussing in detail the subject of fire control, before a group of professional foresters. However, to make it clear how this work has been conducted by us in the interest of wildlife, a brief resume of methods we recommend will be given. In most of our work the lightest of burning will suffice, and this is taken for granted throughout this discussion.

In some controlled burning for quail and wild turkeys, woodland areas under management are cut into small blocks with plowed fire lines, and if necessary special protection is given to certain important thickets, clumps of food plants, and so forth. Then every second or third block, as conditions necessitate, may be burned out in late winter under quiet, slightly damp conditions as after rains, or preferably *at night*.

Need for exceptional economy in fire control as well as in other management measures, has been largely responsible for our development and use of a more economical, and in many ways superior method which

we refer to as "spot burning," though very different from the "spot fires" of forestry nomenclature. Obviously greater care and experience is required in its use as it lacks some of the elements of safety of the method just described.

All burning by this method *must* be conducted on quiet nights with dew fall, else it will not be what we call "spot burning." Open areas in this system require burning out soon after sunset, as dampness develops early there. On the other hand it develops later and to a lesser extent under stands of pine timber, where burning should be started considerably later. Resinous pine needles may even carry a creeping fire throughout the entire night under especially dry conditions, and requires putting out before dew dries in the morning if severe day burns are to be avoided.

Experience with fire indicates that it is highly desirable, even with "spot burning," to divide with fire lines areas to be burned into blocks as small as economically feasible, as a safeguard in case the weather is misjudged. Assuming that this has been done, and that we desire to "spot burn" approximately half the acreage within a block, the procedure is as follows:

Soon after sunset, after appreciable dampness has developed, we experimentally try a fire in a safeguarded corner to see how rapidly the cover burning. If too rapidly, we wait until dampness increases. When judgment indicates that each fire will die out through increasing dampness after it has burned a spot of the desired size, we start criss-crossing the block, setting fires about every one hundred yards until several dozen are burning.

If all goes as planned, the block is ideally "spot burned" when fires die out late in the night, and approximately half of the acreage burned very lightly in well distributed spots, with equally well distributed cover remaining between. If it was too damp when we started, we may have failed to burn the desired proportion, when further burning will be required another night. If we misjudged the weather and a breeze developed, or we set fires too early in old, highly combustible cover, we probably burned more and cleaner than desired. In any case, results will be vastly superior on the average, both from the standpoints of game and timber production, to those secured with uncontrolled day burning.

Now for the procedure to be followed the second and succeeding years, assuming again that we desire to burn about half the acreage. As the "rough" spots left last year will burn with greater intensity than the one year cover, we will start our fires later in the night, and rather centrally in these spots by the method previously described. The fires will die out as they reach the thinner cover (last years' burned areas) , leaving this for the quail and other wildlife. The result is as before, as far as the distribution of burned and unburned areas are concerned, and the older accumulations are removed, and the thin cover left. The burning procedure by this method is similar in general, regardless of the proportion to be burned each year. The greater our experience and the better our judgment, the better results will be, though it should be obvious that *exactitude* cannot be expected by this method. As a rule, it should result in an ideal interspersion of burned and unburned cover, which is what we desire. Needless to say, sufficient help should be available in case plans miscarry, as in all use of fire.

Ticks and chiggers, or "red bugs," constitute a real menace at times to quail, wild turkeys, deer and other wildlife, as well as to man and his domestic animals. In addition game animals frequently have diseases and intestinal parasites which furnish problems in the management of game lands. The speaker holds the belief, and acts upon it at times in his game management work, that properly used fire is of value in the control of certain parasites and diseases, though this has perhaps not been proved by adequate scientific research.

As regards tick and chiggers, the belief is based on practical observations and comparisons over the region under discussion, and is further indicated by the known life histories of the pests in question. For instance, unattached adult female ticks, their eggs in the "duff," and the young "seed ticks" awaiting the passing of host animals, are all presumably vulnerable to fire; so much so that few can escape when it comes. It would seem that they are vulnerable at *all times*, except when actually attached to host animals which can carry them beyond reach of the flames. The same is also true of chiggers, which have a somewhat similar life history.

Theoretically at least, periodic burning or plowing under of ground cover should contribute to quail and wild turkey health by destroying certain disease organisms, as well as the intermediate stages of some of their intestinal parasites. This seems borne out, at least in the case of quail, where large numbers have been examined from environs where the fire history is known. Probably the general soundness of this line of reasoning, will ultimately be proven or disproven by adequate scientific study and experimentation, though it is so difficult that we, at least, have made little headway upon it. Meanwhile we are proceeding in our practical work on the assumption that controlled fire has *some* use here, as well as a *demonstrated value* for purposes previously discussed.

It is not my intention to discuss here the hazard to pine forests occasioned by the fire exclusion policy, the effect of such burning as we advocate upon the fertility, texture, or erosion of soils, or the runoff of water, though all of these matters are being considered by our Association, and taken into consideration in our practical work.

In conclusion, I want to say that I am convinced that forestry and upland game production can frequently go hand in hand, especially perhaps on southeastern longleaf lands. The game manager can so conduct his work as to incidentally grow much valuable pine timber, and those interested in forestry who desire to encourage quail as a by-product of forestry, can produce many of these fine game birds with management practices, including controlled fire of the frequency that now seems desirable for the welfare of the longleaf pine forest.

In both cases, there is need for men thoroughly versed in the use and control of fire, who also know the requirements of the forest, and of the quail, wild turkeys, and other valuable wildlife which inhabit the forest, if intelligent management of these natural resources is to result.

The Place of Fire in Southern Forestry

1942. *Journal of Forestry*. Vol 40.je No. 2. 129–131.

Raymond M. Conarro
U.S. Forest Service

For many years past, investigations have been made, pages have been written, and words have been spoken concerning the use and misuse of fire in the southern forests, the investigations being made by one group opposed to the use of fire, and by another group that believes fire is beneficial. These investigations have been followed by controversial writings, by speeches, by heated arguments. Definite groups have been aligned, always ready to pounce upon, to criticize, and to fly into the press against those of the opposing group. Large fires have burned out of control and the controlled burners have immediately seized upon such a fact as an example of what can be expected; the other group has readily pointed out damage resulting from so-called controlled burning.

The fire exclusionist dwells long and loud upon the inability of a forest soil to produce seedlings where fire is used. The controlled burner likewise shouts from the housetops that a pine forest cannot be established without the use of fire, that pure longleaf stands will ultimately be converted into a climax forest of inferior species. The fire exclusionist immediately replies that more inferior species are found where fire is frequent. On and on goes the battle of words, of investigations, of writings, each group following its own inclinations,

its own views, determined to prove them and still more determined to tell the world about them, and in the meantime our annual southern fire bill still remains high in acreage burned, in damage done, in costs of handling, and John Q. Public becomes more confused, more inclined to follow his own ideas, and yes, much more disgusted with the whole subject. Some forest managers have swung from total exclusion to the use of fire, while others have shifted from the use of fire to total exclusion, and where this has happened the managers seem to be satisfied with their decisions which fortunately or otherwise were largely based upon their own observations and needs and not upon the results of scientific investigations or controversial words. There are other forest managers who have not shifted, who have definite management policies which either require or forbid the use of fire. Again, such policies were largely developed on the basis of their own observations and needs. We have in the South some fine forest properties where fire has been excluded over long periods, also some equally as good where fire· has been used wisely. Likewise, there are forest properties not in good shape where fire has been excluded and also some where burning has been done unwisely. There are plenty of examples of proof for and against fire. One may choose either side and develop a good sound defense of his views.

In addition to these two groups, the exclusionist and the controlled burner, another group must be recognized. This is the group that knows controlled burning is practical and worth while and should be used in managing southern forests. Members of this group fear that to practice their convictions, yes, even to mention doing so, would cause consternation in the ranks of forest managers of other timber types and would spread rapidly the wanton destruction of these types because of the examples these southerners had set. They also fear that controlled burning could not be controlled, that incendiarism would increase, that law enforcement would be less effective, that appropriating bodies would be less impressed when requests are made for fire control appropriations. Yes, these timid souls even fear that the South's forestry program would "go to pot" should controlled burning be acknowledged.

The existence of these three groups, the exclusionist, the controlled burner, and the timid soul, together with the exclusionist crusader and the controlled burning rabble rouser from other than southern parts, with the resulting controversy, the confusion and the fear, have handicapped and delayed considerably an intelligent approach to the real problem of the southern forest land owner, that problem being the profitable management of his woodland.

The agriculture economy of the South and, if you please, the economy of sustained yield forestry demand that the South's forest lands be utilized for purposes in addition to growing trees. Some of these multiple uses conflict with the ideas of the exclusionist, of the controlled burner and of the timid soul, but regardless of that, it must be admitted that many acres of trees are now standing and growing and are being managed which would not be, were it not for

the fact that multiple use makes it financially possible. In plain words, there are many acres of southern forest land that need a crutch, and multiple use supplies just that. This being a fact, should not multiple use be recognized and included as a part of the forest management plan? Such uses as grazing and wildlife production offer opportunities for the land owner to reduce his annual carrying charges and at the same time give more thought and attention to his trees. It is definitely known that both of these uses, to be successful, must have some burned areas. The percentage need not be large or the burn severe; the operation need not be costly, damaging or dangerous. Common sense coupled with knowledge of the property and fire behavior is needed. Fortunately, during the past few years there have been developed instruments designed for the measuring of weather factors which influence fire behavior. By the use of these instruments, the question mark has been removed in determining rates of spread and severity of burn. Along with this development, there have been and still are in progress studies to determine the effects of fire as it is related to weather factors. Fairly accurate knowledge, not complete, but sufficient to chart trends, is now available, which indicates that fire can be used effectively for definite purposes, as, for instance, an area in need of reforestation and which has ample seed trees may be put in much better condition to receive the seed and insure its germination, by the skillful use of fire at the proper time. Likewise, it is known that more effective and economical artificial reforestation can be accomplished, as well as better control of the area during the critical period, if fire is used skillfully. These two examples well may be used to illustrate the subject of this paper. Much as a skilled physician uses his instruments, his training, his skill in a sincere effort to determine the cause of an ailment, the forest land manager should use instruments, his training, and his skill in a sincere effort to determine the cause of a forest condition; then, as does the physician, he should analyze the condition found and prescribe the remedy.

It is known that burns vary with intensity and to the degree that influencing weather conditions vary. It is also known that a burn which would be satisfactory for artificial reforestation would not be at all satisfactory for seed bed preparation, and so on. Each forest condition must be considered separately. There are acres that need the aid that fire can give, while there are other areas which would be damaged by any use of fire. Fire must be considered as a tool, to be used in the practice of forestry in longleaf and slash pine stands and not the means to the end. No farmer would be foolish enough to continue to plow an area after his seed had been sown or to continue to cultivate after the cultivation stage had passed. He would not dust his cotton before the proper period arrived or would he continue to dust after this period had passed. He would not dust his corn in an adjacent field to kill boll weevils but would confine his dusting to cotton. Why then should we use fire before the proper period arrives or after it has passed? Why should we not use fire in a forest type when it can be used effectively, why risk its use in another type when it undoubtedly would be damaging? Why all of this controversy, this confusion, this fear? Are those who manage southern forest lands less intelligent, less skillful, less observing than those who manage southern crop lands? Why not use fire in preparing seed beds, for planting-site preparation, for insects and rodent control, for elimination of plant diseases and weed species, for protection from wild fire, for grazing, for wildlife management, for naval stores operations, and perhaps numerous other reasons? It is known that fire can be used effectively for any of these purposes if the conditions are such as to justify the expenditure necessary to make such use effective. Is there not then a place for fire in the management of our southern forests? The answer is obviously yes, providing it is used strictly in accordance with a plan developed from facts determined by a survey and analysis correlated to available knowledge concerning weather influence and fire effects. If the approach to the use of fire is made in this manner, the forest manager so doing may be likened to the physician who examines his patient, analyzes his findings, and prescribes what is, in his opinion, the remedy. Surely this is a sensible, intelligent, and practicable approach to the use of fire, an approach which can be made by one of average intelligence and which can best be described and understood by the use of the term "Prescribed Burning," burning to a prescription which prescribes the area to be burned, the degree of burn, the method and the time, simple, concise, effective, leaving no room for criticism, for controversy, for misunderstanding. With prescribed burning properly understood and use accordingly, there need be no controversy with the fire exclusionist, no worry for the timid soul that southern practices will be adopted on other forest types to the detriment of such types, or that incendiarism will be spread and law enforcement broken down. There should be no confusion among ourselves, but just the contrary; there should be better understanding, a realization that interests other than trees exist on a high percentage of southern forest lands, and that these interests deserve recognition in accordance with their economic importance. We should no longer consider that fire is 90 or 95 percent, or any other great percentage, of the South's forest problem, but that it is an effective tool, a vehicle upon which sound forestry practices can well rely.

The need for recognition of property rights is apparent. The southern forest land owner has, as does the owner of any property, the inherent right to decide to what use or uses his land will be placed so long as his decisions do not conflict with the public interests. It is not within the province of a stockman, game producer, stumper, or any other, to make this decision. There is no more place for incendiarism, for wild or unprescribed fires in southern pines than there is in any other forest type. To the end that considerable acreage may be saved from this sort of burning, let us be more understanding, more patient, and less timid. Let those with overlapping interests recognize property rights. Let the land owners recognize the economic value of multiple use. Let there be a common understanding between them. It is the obligation of those of the forestry profession to bring about this understanding.

Prescribed Burning versus Public Forest Fire Services

1947. *Journal of Forestry*. Vol 45. No. 11. 804–808.

H. H. Chapman
Professor emeritus, Yale University School of Forestry, New Haven, Conn.: Fellow, S.A.F.

The author recounts the steps which led to the conversion of most foresters to the viewpoint that controlled burning is an essential part of the management of longleaf pine. He disagrees vigorously with those who have not yet adopted this policy.

When the British government organized forestry practice in the various provinces of India, they encountered a number of folk customs which ran counter to the requirements of good silviculture. These folk customs included the two perpetual and closely associated problems of grazing and annual burning. But the Forest Service in India, as organized by Sir Dietrich Brandis, was imbued with the German tradition that exclusion of all fires was the cornerstone of forestry practice.

In 1917, when the writer was with Region 3 of the U.S. Forest Service in the Southwest, he had the pleasure of a three-day visit from a British forester, F. A. Leete, former conservator of the forests of Burma. Mr. Leete described what had happened to Sir Dietrich's no-fire policy in India. About 30 years before Mr. Leete's above visit, a young British forester concluded that fire exclusion was entirely preventing the reproduction of teak, the production of which was a chief objective of management. He said so, and was not only censored but deprived of advancement and thoroughly squelched. Shortly before Mr. Leete left Burma 20 years later, a detailed field examination, participated in by Leete, confirmed the young forester's statements. The Forest Service personnel could have stuck their heads in the sand, but they wanted to succeed in growing teak. From then on, fire played its role in British management of Indian forests[1] and was soon applied also to the Chir pine in the Himalayas.

The U. S. Forest Service, before it had acquired any forests or any land-management

---

1. Chapman, H.H. The use of fire in regeneration of certain types of forest in India. Jour. Forestry 25:92–94. 1927.

experience in the South, was guided only by its knowledge of the destructiveness of fire throughout the Northeast, Lake States, Mountain States, Pacific Coast, and California. Consequently it dug equally solid foundations for the structure of total exclusion of fire, which it later extended to the new national forests in the southern pines. Instruction in all American schools of forestry was in harmony with this position. Foresters who for any reason flirted publicly with fire as a silvicultural tool were severely and sincerely condemned. The patient and laborious effort required to awaken the public to the evil effects of forest fires and to get the situation under control bulked extremely large. Consequently the initial reaction of the profession as a whole was to reject even the sanctioning of investigations into this subject, not to speak of the publication of results, in the belief that more harm than good must come of inevitable misunderstanding and misinterpretation of possible benefits from prescribed use of fire. This impression was strengthened by a premature and uninformed popular campaign for light burning in California in the decade 1910–1920.

The writer first began his observations on the ecology of southern pines in the Missouri Ozarks in 1907, where fires were evidently the cause of destruction of practically all small, shaded seedling shortleaf pines on uncut areas. But in 1908, in the longleaf pine hills of north central Alabama, the absolute necessity of a ground fire to expose mineral soil was demonstrated on contrasted areas where the seed crop was so large that every worm hole in the rotten logs contained a sprouting seedling pine! At that time the writer concluded that close observation of ecological behavior must be

substituted for precepts inculcated in the classroom and in existing practice, and that for a given species and region, the behavior of the species and its reaction to environment might not conform to that of other species in distant regions with different environment. This seemed like a sensible view in the light of the fact that man proposes but nature disposes, and that only by conforming with natural laws can he turn them to his benefit—as he has done in agriculture.

In 1912 in *American Forestry*[2] the writer stated regarding prevailing burning practices in the South, "The effect of these fires upon the forest has been deplored by foresters, and the tendency seems to be to try to pass laws modeled after those of northern states, which seek to absolutely prevent fire in the forests and establish a system of fire wardens for the purpose. But it is more than possible that such a policy in the South would defeat its own ends and should never be attempted.[3] It is the right policy in the northern states, where fires can and should be absolutely prevented. But there is abundant evidence that the attempt to keep fire entirely out of southern pine lands might finally result in complete destruction of the [pine] forests... The risk gets worse as the period extends, till at the end of 10 to 15 years, if fire is set in a dry time, the mature longleaf timber may be killed. [Many years later this prophecy was abundantly fulfilled.]

"Shortleaf [and loblolly] seedlings are very easily destroyed by fire. But the young trees soon develop a thick bark and will resist small ground fires. In a region studied this spring in Arkansas [Crossett], it was found that it took the average seedling only five years to reach a diameter of over an inch, and become fairly fire resistant...

---

2. Chapman, H.H. Forest fires and forestry in the southern states. Amer. Forestry 18:510–517. 1912.

3. The reader should complete the reading of this article before assuming that these early opinions as to organized fire service are still "held."

Seedlings growing in the forest under partial shade grow more slowly and may be killed by fire at 8 or 10 years…. The proper use of fire, and not complete fire prevention, is the only solution of the problem of future forestry in the South." (This was one basis for the writer's later advocacy of prescribed fire for loblolly pine.)[4]

Our investigation of the ecological role of fire in securing longleaf pine reproduction and survival was fortunately located permanently at Urania, Louisiana from 1920 to date, on the land of Henry Hardtner, to whose willingness to cooperate in actually applying the technique of prescribed fire and taking the risk, on his own property, the South and the profession owe a debt which can never be repaid. By 1926 it had been learned that a fire preceding the seedfall was only the first step. Exclusion of subsequent fires had the following results:

1. Prevented all subsequent establishment of longleaf seedlings.
2. Permitted competing species of pines and hardwood to exterminate established longleaf seedlings by shade.
3. Most important of all, permitted brown spot disease to defoliate, stunt, and kill the seedlings and decimate the stand.

Later, in 1932–34, the final demonstration took place—complete loss of nine-tenths of the 1913 pole stand of pure longleaf pine, through four successive, uncontrolled summer and early fall fires on land unburned for from 19 to 21 years.

In 1926 a fire rotation of three years was stated as the controlling ecological factor for survival of the species, under natural reproduction from seed.[5] This started an American cycle of nearly twenty years, following the British precedent, which was required before professional opinion, both public and private, was converted to use of prescribed fire for longleaf pine. A similar process of incubation is still going on with reference to loblolly pine in the South, and pitch pine in the Northeast.

In sketching the historical development of this process of conversion to and acceptance of an idea so revolutionary and antagonistic to universally accepted precepts of silvicultural practice, it must be borne in mind that new and different ideas must usually be first advanced by individuals and will stand or fall, in the modern scientific era, only by the accumulation of facts based on the two parallel processes of observation or experience, on the one hand, and controlled experimental research, on the other. The weight of universally accepted belief cannot be set aside or overcome merely by announcing a new theory. When existing tenets are reinforced by official regulation and discipline, it takes many years of constant agitation, based on facts, to permeate the mass of scientific opinion, and a further period to conquer the last stronghold of administrative resistance.

As was to be expected, the initial impetus came from private (in this case institutional) research in cooperation with the equally untrammeled initiative of private industry. No doubt *individuals* in the employ of government or states may have sensed the situation just as soon, as was the case in the Southwest with regard to grazing damage (1916–1918). But in a large organization the individual is obviously and necessarily governed by accepted principles and policy and cannot publicly discuss or debate new policies until the governing body has officially accepted them. The example of the

4.   Chapman, H.H. Management of loblolly pine in the pine-hardwood region in Arkansas and in Louisiana west of the Mississippi River. Bul. 49, Yale School of Forestry. 1942.
5.   Chapman, H.H. Factors determining the reproduction of longleaf pine on cutover lands in LaSalle Parish, La. Yale School of Forestry Bul. 16. 1926.

young British forester in India is a case in point.

The conversion of the federal authorities to the new practice of prescribed fire for longleaf pine was the result of three separate influences; first, the publication of facts in the *Journal of Forestry*, whose columns are open for discussion of all phases of forestry—the most important function of a professional journal; second, actual experience by public employees on the ground, in managing national forests (in Florida and Texas) for longleaf pine reproduction by seed; third, and most important, research conducted by the Southern Forest Experiment Station, which had its origin according to E. L. Demmon, director, in 1932, or six years after the publication of Bulletin 26 by the Yale School of Forestry.

During the period 1931–1944, 26 bulletins and articles were published by this station dealing with fire in the longleaf and slash pine types. The two published in 1931 as notes in the *Forest Worker* dealt solely with damage done by uncontrolled fires. Later articles by various authors dealt with control of brown spot by fire (Siggers), the effect of fire on soils and on stand composition (Heyward), administrative problems of fire control (Eldredge), methods of controlling fire (Bruce, Bickford), silvicultural aspects of fire in longleaf pine type (Demmon), and others. From these studies there emerged the body of technical information which established the facts previously assumed in Bulletin 26, and additional important information to the effect, for instance, that burned land contained more nitrogen than unburned areas, due to the fact that grass roots constituted the chief source of humus. In this decade, 1920–1930, a 10-year experiment at McNeill, Mississippi was conducted by the Southern Forest Experiment Station in cooperation with the Bureau of Animal Husbandry to test the effect of annual fires versus total exclusion of fire on longleaf pine reproduction and cattle grazing. This experiment was planned before anyone had advocated the use of prescribed fire at 3-year intervals and was concluded without modification of the original plan. Meanwhile, S. W. Greene, the animal husbandry cooperator, basing his conclusions on the results obtained, had shown that cattle put on appreciably more weight on burned than on unburned areas, a fact which was vigorously combatted by foresters in general, in defiance of the evidence. The mistake Greene made was in claiming that *annual* fires were beneficial to longleaf pine re· production. The writer conducted a vigorous private correspondence with him on this subject, being convinced that annual fires were completely destructive to annual seedlings.

In 1939 Wahlenberg published the results of this experiment (co-authors S. W. Greene and H. R. Reed).[6] He showed that annual fires and total exclusion of fire were equally destructive to longleaf pine reproduction. But he went further. No provision had been made in the experiment in 1920 for testing the effect of fire at 2- to 3-year intervals nor, as stated, was any modification made in this respect during its execution. But on adjoining land fires had burned at irregular, not annual, intervals. He found that this had resulted in a normal growth of longleaf pine seedlings and incorporated this result in the manuscript.

Then came the assault on the final stronghold, the administrative officers, who were in last analysis responsible for the effects of publishing officially any advocacy of the use of fire for any purpose. Their attitude, and in the light of their responsibilities no one could criticise them for it, was shown

---

6.   Wahlenberg, W.G. Effects of fire and cattle grazing on longleaf pine lands as studies at McNeill, Miss. U.S.D.A. Tech. Bul. 683, 52 pp. 1939.

at the session of the Society of American Foresters in December, 1934 at Washington, D. C. in a program arranged by E. L. Demmon, director of the Southern Forest Experiment Station, which placed on record some of the conclusions arrived at to the effect that prescribed fire had its place in southern silviculture.[7] Roy Headley, chief of Fire Control, announced determined opposition to the official acceptance of this idea as a rule of practice, fearing its abuse by uninformed and irresponsible persons. Wahlenberg's bulletin was held from publication for six years while the battle raged between research and administration. It appeared finally in 1939. This was the first official recognition by Washington of the advisability of the use of prescribed fire in longleaf pine forests.

Meanwhile, some very serious wild fires occurred after protection on large areas belonging to lumber companies and others, who were trying to grow timber. Among those were the destructive 15,000-acre fire on the property of Alex Sessoms in southern Georgia, resulting directly from fire exclusion for 10 or more years, and a similar 5,000-acre fire on the Jackson Lumber Company's lands in southern Alabama. The lesson of these fires, and the exercise of the faculties of observation and reasoning, started the private practice of prescribed burning even before its final acceptance by the federal authorities. An outstanding instance was that of the Superior Pine Products Company at Fargo, Georgia, on 200,000 acres, burned on a three-year rotation, for both slash and longleaf pine, since about 1940. Other firms, however, had still adhered to the initial advice of foresters and were continuing the policy of fire exclusion for the time being.

The Jackson Lumber Company of Lockhart, Alabama, after a few experimental burns in 1944–45, began systematic prescribed burning in 1945–46 and are continuing this policy. Their 80,000 acres of cutover longleaf pine land had been largely reproduced to longleaf pine because of their previous efforts, from 1924 on, to exclude fire which had succeeded in reducing the fire frequency from an annual to an occasional, approximately 3-year basis. The areas so burned were the ones on which healthy reproduction had become established.

With the precedent of Wahlenberg's bulletin, backed by accumulating experience on the Florida, Mississippi, Louisiana, and Texas national forests, the Forest Service administrative officials in both Atlanta, Georgia, and Washington, D.C., soon came to the position not only of practicing prescribed fire on the national forests for longleaf pine, but also permitted the publication, in 1942, of an article in the JOURNAL OF FORESTRY on the use of fire in the flatwoods of the South,[8] and finally, official instructions to the general landowning public in the use of fire for protection of forests of longleaf and slash pine.[9]

The last stronghold of intrenched opposition to prescribed fire for longleaf pine lies in the state forest services of some of the southern states—we will not specify which or to what extent. The basis for this position is in one respect sound. Prescribed fire will never even diminish, much less abolish, the need for ever-increasing efficiency in state fire protection. The season for safe burning is confined to late December, January, and February. Safe burning also requires attention to conditions of ground moisture, relative humidity, wind movement, condition of the rough, and forest type. These

---

7.   See Jour. Forestry 33:320–337. 1935.
8.   Bickford, C.A. The use of fire on the flatwoods of the South. Jour. Forestry 40:973. 1942.
9.   Bickford, C.A. and J.R. Curry. The use of fire in the protection of longleaf and slash pine forests. South For. Exp. Sta. Occas. Paper 105, 22p. 1943.

requirements necessitate trained and constant supervision if damage is to be reduced to negligible proportions and if beneficial results are to be secured. Fires at any other season, or under bad conditions, or with improper management, even in the favorable months, not only damage but may completely destroy stands even of mature timber. But abundant evidence exists that frequent (3-year) prescribed fire may save a stand from destruction even in an off season. It has done so at Urania, Louisiana. Hence there is no conflict between prescribed fire and eternal vigilance to prevent wild fire anywhere, any time. The two go hand in hand.

The second objection is more serious, but in my belief less valid. It constitutes a fundamental lack of trust in the innate intelligence of farm and forest owners and workers, in the belief that any tolerance of fire in the forest will cause them to revert to the policy of uncontrolled annual burning with all of its destructive consequences and cause the situation to get completely out of hand. Rather than start something that they believe cannot be controlled, certain state forest services prefer to close their eyes tight and adhere to the line of absolute exclusion of fire regardless of the rising tide of information and knowledge of proper silvicultural practices. This attitude was responsible for the dismissal of one southern state forester.

The writer takes a diametrically opposite viewpoint as to the ability of the southern rural population to absorb facts regarding fire. They know too much already, from daily and lifelong contact with actual conditions in the woods. They do need further guidance in the effects of annual fires in ruining all chances for reproduction and even impoverishing the grazing. Is this task of education too formidable for state men to tackle, when its accomplishment means the future prosperity of practically the entire forested pine regions of the South. The

answer is that it has not been tried. In the long run such a policy of deliberately repudiating or concealing known facts is bound to fail and to bring permanent discredit upon those who advocate it and upon the profession which they represent. It is unsound scientifically, professionally, and from the standpoint of psychology and public cooperation. The writer recently prepared a short pamphlet for farmers in cooperation with the Jackson Lumber Company at Lockhart, Alabama, of which several thousand copies were distributed. The results were all that were hoped. Cooperation has increased amazingly, along with respect for the "intelligence" of the advocates of prescribed fire. "Those professors finally found out what we've known all the time" (ignoring some things they had not known). The same result was obtained by the Superior Pine Products Company of Georgia.

Fire as a dual purpose tool for protection and silviculture has a limited use in other parts of the country, and is applicable only in certain types. Of these, an example is the pitch-shortleaf pine-sprout-hardwood type, extending from New Jersey to the Cape Cod area of Massachusetts, and the white pine type in Idaho. The facts regarding the proper regime for prescribed fire, both in reducing the chance for repeated conflagrations and restoring the pitch and shortleaf pines to their former dominance in the type, have already been demonstrated by a 10-year cooperative experiment between the U. S. Forest Service and the State Conservation Commission of New Jersey.

What will be the future attitude of the state forest fire services upon which rest the absolute necessity of eliminating recurrent holocausts? Wild fires in New Jersey have reduced a huge area of pine barrens to an unproductive state which it will take 20 to 30 years of fire control to restore. And how about Cape Cod? Forest fire control by state agencies, and prescribed burning in certain types of forest to reduce fire hazards and bring hack the profitable species of timber, are Siamese twins. The one cannot continue to thrive without the other, and quarrels between Siamese twins are both uncomfortable and unprofitable.

Prescribed Burning

1975. *Journal of Forestry*. Vol 73. No 12. 776–780.

Robert W. Cooper

Research Forester, Southern Forest Fire Laboratory, Forest Service, U. S. Department of Agriculture, Macon, Georgia

Abstract: Prescribed burning is a preferred treatment in many fuel management situations because of its low cost, compatibility with other land-use objectives, and little or no undesirable side effects. The problems, limitations, and associated consequences of fire treatments are evaluated and compared with alternatives.

Fuels are an inherent part of the forest community. Unless we're willing to live with the threat of disastrous wildfires, fuel management becomes a must. Without some form of fuel management, wildfires are inevitable. If not today, then tomorrow! Prescribed burning, the intentional application of fire to forest land under selected conditions, is one form of fuel management that offers a feasible solution to this problem.

In the South, this management tool has already been put to the test and has proved itself time and time again. More than 2 million forest acres are treated each year with a fire prescription. To evaluate the effectiveness of prescribed burning as a means of fuel management—i.e., in reducing the number, size, and intensity of wildfires—a study in the southern Coastal Plains tested the relation between understory fuel accumulations (rough) and wildfires (4). Results indicate clearly that the large, destructive wildfires occurred, in almost every case, in heavy fuels where prescription burning was not practiced. Annual burn percents during 4 years ranged from 0.03 percent in the youngest roughs to 0.14 percent

in the 5-year-old roughs, and an unexpected 7.00 percent in roughs that had never been treated with fire. Height of stem bark char, a good indication of intensity, averaged about 2 feet in roughs less than 2 years old, compared with about 20 feet in roughs 5 years and older. There is little doubt that prescribed burning has functioned as a most effective means of fuel management in southern forests. It continues to do so today.

Application—When and How

In the South, hazard reduction burns are generally carried out during the dormant season when: air temperature is low (less than 50° F.); upper litter moisture is relatively low (8 to 12 percent) and lower litter moisture moderately high (20+ percent); relative humidity is between 20 and 50 percent; and winds are steady (2). On most pine sites, natural fuels build back to a critical level in about 5 years. Repeat burning is recommended to keep the fuels in check.

Slash disposal burning can be accomplished during any month of the year if the fuel conditions are right and favorable weather prevails. In summer, logging debris reaches its optimum combustion level about

Figure 1. Prescribed fire is a practical, inexpensive form of fuel management.

2 months after cutting. In winter, about 4 months are required to achieve maximum combustion efficiency (8).

In the West, most prescribed burning is scheduled for the fall months (1). Snow cover generally prevails during the winter season on the higher northern sites, precluding any fire activity. Although a few days or weeks in May and June are sometimes suitable for slash or brush broadcast burning, the spring and summer months bring extended periods of hot, dry weather and the major wildfire season. Nevertheless, summer burning is presently finding increased popularity along the Pacific coastline. Although still considered risky business, fire prescriptions offer the best solution to the fuel problem. The mote precarious burning chances may warrant a delay until the fall season. In the Pacific Northwest, moist, oceanic air-masses generally move in early in September; burning of hold-over blocks can begin after the first general rain. Those higher elevation blocks with north and east aspects are broadcast burned first; those on lower south- and west-facing slopes are burned later. The burning of piles and windrows is best deferred until later in the autumn.

For the most part, westerners prefer burning conditions not too divergent from those appropriate in the East. Fuel moisture contents between 6 and 15 percent are acceptable; relative humidities from 20 to 50 percent are suitable; wind speeds of 2 to 10 mph are preferred. In broadcast burning, the stronger winds are desired because they are generally more persistent and less likely to switch in direction. Lighter winds are often acceptable when burning piles and windrows.

Figure 2. Understory vegetation and needle fall combine to produce a hazardous fuel condition in many southern pine stands.

Backfiring (forcing the fire to advance into the wind) is the most common technique used in hazard reduction burning, but strip, spot, or flank firing may be appropriate in special situations. Although backfires are relatively expensive, they are easy to manage and control, and are generally most effective in reducing fuel accumulations to an acceptable level. A well executed prescribed backfire can consume about two-thirds of the understory fuel without damage to the soil, overstory, or other amenities of the forest. In fact, in most instances, it is the only feasible means of applying fire to young stands where the overstory is less than 30 feet taller than the understory. The initial burn in heavy fuel accumulations is almost always a backfire because of the associated safety factor. Successive burns, of course, can be carried out with strip heads, spots, flanks, and other less expensive techniques.

In the South, fire sets are usually activated by hand (drip torches). Control lines consist of natural barriers, roads, and plowed firebreaks. Topographic features are generally not serious factors to contend with since most operations are limited to flat or rolling terrain. In the West, topography is an important consideration in selecting ignition patterns and regulating fire behavior. Fires generally head upslope and back downslope. Coves, ridgetops, bottoms, and slope thermal belts all have an effect on behavior. Because many of the blocks are relatively inaccessible, remote ignition offers many advantages. In heavy fuels, special ignition devices may be required to assure rapid and persistent fire starts. Firelines are usually constructed with bulldozers; a 12-foot-wide break is considered minimal. Fuel modification or manipulation before burning is a common practice in many western operations.

Pros and Cons of Fire Treatments

Fuel management is more than just hazard control. It includes the modification of understory fuels to improve site and seedbed conditions for regeneration; it includes the

Figure 3. Smoke from forest fuel combustion (particularly headfire) may cause problems in smoke sensitive areas.

enhancement of wildlife habitats; it includes the control of vegetative fuels to promote the growth of crop trees; it includes the magnification of the forest environment for recreation and other amenity values. Fire is capable of accomplishing more of these objectives than other fuel management treatments.

Alternative treatments cannot compete with fire in practicality and cost. Average costs in the South are still less than $1 per acre, although costs as high as $5 per acre have been reported for some individual burns. In the West, costs are considerably higher-ranging in general from $1 to $50 per acre, with occasional costs as high as $100 per acre. Yet, when compared with the· costs of chemical or mechanical treatment—$25 to $300 per acre—prescribed burning is still a bargain.

In addition to its low cost, it is a preferred treatment in many fuel management situations because it is compatible with other land-use objectives, is less destructive to the site than other remedies, and produces a minimum of undesirable side effects. Investment costs are relatively low. The general appearance and accessibility of treated areas are enhanced considerably. The ecological makeup of forests may be disturbed more by fire exclusion than by judicious use, for fire is a natural part of the environment. Without it, foresters would have to contend with unbearable management costs, intolerable fuel situations, and a general decline in the productivity of natural resources.

On the other side of the coin, we find that there may be inherent shortcomings in fire prescriptions. Single burns are seldom adequate. As a general rule, a series of treatments is necessary during the life of a stand to achieve fuel management objectives. In addition, there is always the chance that a prescribed fire might escape and become a wildfire. Until such time as weather conditions can be precisely predicted, fire prescriptions possess an unusual degree of risk. Ideal burning days are often few and far between. Coupled with this fact is the reality that many of the suitable days are not predictable and, consequently, not used.

Some temporary impairment of the atmosphere, at the least, may occur as a result of prescribed burning. Smoke from forest fuel combustion may create problems along major travel routes and nearby communities, or contribute to an already critical air pollution episode over a specific geographic area.

Finally, one might question the wisdom of wasting a potential source of energy when the nation is faced with a critical shortage. As much as 20 million tons of forest fuel may be going up in smoke each year from prescribed burning operations. If it were economically feasible and practical to harvest, use, and market understory forest fuel, in excess of 1 billion tons might be made available annually on a sustained yield basis.

Conclusions

Prescribed fire, once considered an art in itself, has achieved scientific status and the potential to play an ever-increasing role in the management of forest resources. If fuels are to be managed, fire must be considered as one of the more practical tools at our disposal.

We've identified the fuel and weather conditions necessary for a successful treatment (2); we've developed firing techniques and practices that produce required intensities and behavior (6); we can predict the effects and responses from various burning operations (5). All things considered, we've almost mastered this tremendous force. We've reached the summit—only to find that new challenges confront us.

Although, to my knowledge, there is no evidence to indicate that air quality is

Figure 4. Backfire smoke is dissipated rapidly as it spreads downwind.

Figure 5. A good prescribed burn. Fuel accumulations are reduced drastically without damage to the overstory crop trees.

permanently impaired in areas where prescribed burning is practiced, there is clamor and question regarding the environmental effects of fire treatments. As foresters, we are as concerned about the environment as the next man—more so I hope. Smokes may look the same, but they seldom are. The major products of forest fuel combustion are particulates, carbon dioxide, water, carbon monoxide, and certain hydrocarbons. Most of what we actually see are the particulates and water vapor. One area of concern deals with the hydrocarbons and carbon monoxide, both of which result from insufficient oxidation or inefficient combustion processes. Oxidation processes in the upper atmosphere, which receive shortwave ultraviolet radiation from the sun, convert carbon monoxide to the dioxide with time

(7). Most evidence indicates that nearly all of the particulates and many of the gases adsorbed on their surf ace are washed out by precipitation. Consequently, it is reasonable to assume that most prescribed fire effluents that remain suspended in the lower atmosphere are short-lived. They may, in fact, actually be responsible for the washout. Smoke particles act as condensation nuclei that start precipitation; soluble gases are dissolved in rainfall and in the oceans; particulates are washed out or fall out due to wind and gravity (3). Indeed, the air has a great capacity for cleansing itself

Another area of concern deals with smoke management. Prescribed fires have been responsible for reduced visibility and dangerous traffic situations on major highways and expressways as a result of persistent smoke. Communities and centers of population located downwind are particularly vulnerable. On occasion, prescribed fires escape. The potential benefits of the prescription may be nullified. We have been known to carry out a burn in order to satisfy a schedule or fulfill an allotment. In most of these situations, our problem is one of poor judgment. Nevertheless, we must learn how to fulfill the objectives of a sound prescription without creating visibility and nuisance aggravations from smoke.

Prescribed fire is capable of minimizing fuel hazards while accomplishing other objectives of forest resource management. It has become an indispensable tool of management in softwood forests of the West, as well as the pinelands of the South. Whether fuel management involves the reduction of heavy accumulations of logging debris, the control of undesirable understory vegetation, the removal of deep litter layers to permit regeneration, or the minimizing of natural fuel hazards in the forest, fire fits into the picture. Most foresters view it as a scientific weapon applicable to many situations and ailments of the forest community. They also recognize that circumstances exist when fire is neither desirable nor needed.

There are, of course, alternatives that might serve the same purpose. Fuel utilization might eventually be the best answer to fuel management if it becomes economically feasible. For the present, however, all things considered, alternative treatments are generally more costly, incur more undesirable side effects, and seldom exhibit the diversity and multiple-purpose achievements of prescribed fire.

Literature Cited

1. Beaufait, W.R. 1966. Prescribed fire planning in the Intermountain West, USDA Forest Serv. Res. Pap. INT-26, 27 pp. Ogden, Utah.

2. Cooper, R.W. Knowing when to burn. Second Annu. Tall Timbers Fire Ecol. Conf. Proc.: 31–34.

3. Cramer, O.P. 1969. Disposal of logging residues without damage to air quality. Weather Bur. Tech. Mem. WR-37, U.S. Dep. Commerce, 8 pp.

4. Davis, L., and R.W. Cooper. 1963. How prescribed burning affects wildfire occurrence. J. Forestry 61: 915–917.

5. Hough, W.A. 1968. Fuel consumption and fire behavior of hazard reduction burns, USDA Forest Serv. Res. Pap. SE-36, 7 pp. Asheville, N.C.

6. Mobley, H.E., et al. 1973. A guide for prescribed fire in southern forests. Southeast. Area, State & Priv. Forestry-2, USDA Forest Serv., 40 pp. Atlanta, Ga.

7. Stern, A.C. 1968. Air pollution and its effects. Acad. Press, New York–London, 694 pp.

8. Wade, D.D. 1973. Logging residues disposal hazards in Georgia's Piedmont. South. Lumberman 226(2802): 15–18.

Crown-Fire Potential in a Sequoia Forest after Prescribed Burning

1975. *Forest Science*. Vol 21. No. 1. 83–87.

Bruce M. Kilgore and Rodney W. Sando

The authors are, respectively, Associate Regional Director, Professional Services, Western Region, National Park Service, San Francisco, and Instructor, College of Forestry, University of Minnesota, St. Paul.

Abstract: Prescribed burning in a giant sequoia-mixed conifer forest reduced the potential for high intensity surface fires and crown fires. Based on three burned plots and three control plots, fuels on the ground were reduced from 203.5 to 30.1 tonnes/ha, while live crown fuels were reduced from 18.0 to 7.8 tonnes/ha. The lowest segment of the forest crown having more than 45.4 kg of fuel per 30.5 cm per 0.4 ha increased from about 0.9 to 4.9 m. Fuel complex porosity for the unburned plots indicates little chance of a sweeping crown fire independent of a surface fire, but heavy ground fuels can support a fire of intensity sufficient to bum individual tree crowns. Under preburn fuel and severe weather conditions, a wildland fire spread model predicted a forward rate of spread of 3.8 cm/sec for a ground fire and a reaction intensity of 28.8 cal/cm/sec. After prescribed burning, the same model predicted a spread rate of 0.05 cm/sec and a reaction intensity of 0.25 cal/ cm/sec. Surface fuels accumulate again rapidly. The longer term impact of prescribed burning was killing smaller trees and the lower levels of live crowns in larger trees, thus removing fuel from the intermediate layer between surface and crown fuels.

Keywords: Fire spread, fuel reduction, crown fuels

Crown fire is a major threat to the continued existence of a giant sequoia-mixed conifer forest. The probability of a surface fire moving into the crowns of giant sequoia (*Sequoiadendron giganteum* [Lindi.] Buchh.) in a given mixed conifer forest is increased by an understory of young saplings—particularly white fir (*Abies concolor* [Bord. & Glend.] Lindl.). With their low-hanging branches, such saplings form an almost continuous fuel supply from ground level to their tops— 3 to 15.3 m (10 to 50 ft) high. Frequently there are thickets beneath larger trees which extend to 30.5 m (100 ft) or more in height; these larger trees frequently reach the lower crown of mature trees (from 54.9 to 76.3 m or 180 to 250 ft tall). This ladderlike arrangement of fuels creates a high potential for a fire to pass from surface fuels into sequoia crowns. Unlike its close relative, the coast redwood (*Sequoia sempervirons* [D. Don] Endl.), the giant sequoia will not sprout; hence, if all needles are killed by crown fire, the tree will die.

In the past, small undergrowth trees were readily killed by light surface fires which regularly moved through these forests (Kilgore 1973b). A continuous fuel layer from ground to crown was thus largely prevented by the normal fire regime found in most of the Sierra Nevada mixed conifer forest. More effective wildfire suppression programs have prevented light surface fires which would have eliminated all but a few young trees. Thus the fuel complex has changed (Biswell *et al.* 1968, Wilson and Dell 1971) and the potential for crown fire has increased.

It is important to distinguish two different forms of forest fires commonly referred to as crown fires. The first and most destructive occurs when fuel and weather allow fire to spread between individual tree crowns—either with or without surface burning. The second occurs when tree crowns burn individually and is commonly termed "crowning out" or "torching" (Davis 1959). Both forms pose a threat to the sequoia overstory.

Crown-fire potential is a function of the quantity and arrangement of fuels. Surface fuels play an important role in the development of crown fires. Van Wagner (1968a) concluded that crown fires in red pine plantations are extremely unlikely unless supported by fires in lower strata. On the other hand, fuel types with certain physical or chemical characteristics have been known to support crown fires independent of surface fires under extreme environmental conditions, usually including strong winds. The majority of crown fires, including those m sequoia-mixed conifer forests, generally burn in conjunction with surface fires.

The reduction of crown-fire potential in the sequoia forest may be accomplished by removing some surface fuels or intermediate level and crown fuels up to some "safe" level not ignitable by fire in surface fuels. This may be accomplished by low intensity prescribed burning under weather and firing techniques that prevent crown fires.

The purpose of this study was to determine how the potential for crown fires in a sequoia-mixed conifer forest was affected by the moderately intense prescribed burning conducted in November 1970, at Redwood Mountain, Kings Canyon National Park, California (Kilgore 197 3 a). The study area, the methods employed, and the results obtained, are described together with a discussion of the short-term and longterm potential for crown fires following prescribed burning.

Study Area and Methods

Data related to crown fire potential were collected in 1971 and 1972 on six of the 18.3 x 30.5-m (60 × 100-ft) study plots described in earlier work (Kilgore 1973a). Three were control plots and three plots had been burned in the November 1970 prescribed burn. The plots were slightly east of the ridge of Redwood Mountain, Kings Canyon National Park, California, at 2013 m (6600 ft) elevation. The climate, soils, vegetation, and previous human impact on the area were described in the earlier work. In general, this largest grove of giant sequoias includes the usual southern Sierra mixed-conifer forest species, namely white fir, sugar pine (*Pinus lambertiana* Dougl.), ponderosa pine (*Pinus ponderosa* Laws.), and some incense-cedar (*Libocedrus decurrens* Torr.) and California black oak (*Quercus kelloggii* Newb.).

The method of describing the crown fuel complex was that of Sando and Wick (1972) in which both weight and volume of crown fuel and its physical location are determined. Trees in two diameter categories—more than and less than 2.5 cm (1.0 in) dbh—were measured on six 0.004-ha (0.01- acre) sub-plots systematically located within each macroplot; mature giant sequoia were not included. Each tree was identified by species, and measurements were made of diameter, height, and crown length and crown width. These measurements were used to supply data for species regression equations that predict oven-dry weight of crown material (Storey *et al.* 1955, Fahnestock 1960). The final weight values were apportioned depending on the physical dimensions of each tree and then graphed by computer (Fig. 1), thus giving a graphical representation of the vertical distribution of live crown materials less than 6.4- cm (2.5-in) diameter. The assumptions involved in this method are noted in Sando and Wick (1972).

In order to convert these data from a graphical form to an integrated quantitative form, two indices are used: the crown volume ratio (CVR) and the mean height of crown base (MHCB). CVR is defined as the ratio of the total space from ground to tree-top level to the space occupied by the live tree crowns themselves. MHCB is defined as the height below which no 0.3-m (1-ft) interval has more than 112.4 kg/ha (100 lbs/acre) of crown fuel.

Surface fuels were sampled 8 months after the November 1970 fire using the line intersect method (Van Wagner 1968b). Some post-fire accumulation had already occurred. Eight transects, 7.6 m (25 ft) long, were run in each macroplot. The diameter of each piece of fuel more than 6.4 cm (2.5 in) in diameter was recorded along all of the transects, while the smaller-sized fuels were measured only on the last 0.9 m (3 ft) of each transect. Small-sized surface fuels were grouped into three classes: (1) less than 0.64 cm (0.25 in); (2) 0.64 to 1.9 cm (0.75 in); and (3) 1.9 to 6.4 cm (2.5 in). Average diameters for each class are presented in Table 1. Larger surface fuels were put in the nearest 2.5-cm (1-in) diameter

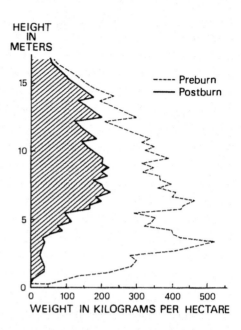

Figure 1. Vertical distribution of oven-dry crown weight before and after burning in a giant sequoia-mixed conifer forest.

class. Depth and weight of duff were already available (Kilgore 1973a).

Results

The fire burned on November 23–24, 1970, when the air temperature was 14.4 °C (58°F), relative humidity was 20 percent, fuel moisture levels were 10 percent, and there was essentially no wind (Kilgore 1973a). Before burning, the forest floor was covered with a large volume of dead fuel. The surface litter layer was well developed from needles and small woody material. In addition, large fallen logs were abundant. The duff layer was also well developed and sometimes over 15.3 cm (6 in) deep. The weight of these surface fuels totaled 203.5 tonnes/ha (81 tons/ acre) before burning (Table 1). These fuels accumulated in the absence of fire and provided ample fuel for an intense surface fire.

Live crown fuels in the lower 16.8 m (55 ft) of the stand were also abundant

Table 1. Weights of ground fuels before and after prescribed burning in a giant sequoia-mixed conifer forest.

	Preburn			Postburn		
Fuel	Weight per hectare	S.D.	SE$_{\bar{x}}$	Weight per hectare	S.D.	SE$_{\bar{x}}$
	Tonnes			Tonnes		
Litter	3.8	1.5	0.2	0.8	1.0	0.1
Duff	94.4	50.1	3.2	11.7	14.7	1.0
Class I[1]	1.2	.9	.1	.4	.3	([2])
Class II[1]	2.4	1.7	.1	.5	1.1	.1
Class III[1]	4.2	7.3	.6	1.7	5.0	.4
Greater than 6.4 cm	97.5			15.0		
Total	203.5			30.1		

[1] Mean diameters used for particle size classes are as follows: Class I, 0.38 cm; Class II, 1.22 cm; Class III, 3.51 cm.
[2] Less than 0.05.

before burning (Table 2, Fig. 1). The base of the live crown fuel complex was close to the ground as represented by the MHCB.

Following the fire, the fuels were substantially reduced (Tables 1 and 2). Much of this change involved heavy ground fuels and duff which burned slowly after the main fire front had passed. Nevertheless, live crown fuels in the lower canopy were reduced more than half, and the MHCB increased substantially (Fig. 1 and Table 2).

Discussion

The impact of the prescribed burning on the live crown fuels is difficult to assess. The CVR was 108 before burning and 142 after burning, reflecting the open character of the fuel complex. By comparison, red pine plantations and jack pine stands described by Sando and Wick (1972) had CVR's of 10.5 and 3.0 respectively. The packing ratio (B) or ratio of actual fuel volume to maximum possible fuel volume may be used to express fuel bed porosity (Countryman and Philpot 1970). We calculated the live crown packing ratio for each 1.52-m (5-ft) stratum in the lower 15.2 m (50 ft) of the stand before and after burning (Table 3). The low B values at all levels reflect the relatively low crown weights found in this stand.

Porosity values for the unburned stand and the relatively high CVR indicate there is probably little likelihood of a sweeping crown fire in this stand independent of a fire in surface fuels. However, surface fuels will support a fire of considerable intensity which might cause "torching" of individual tree crowns. Because of the restricted range and great age of giant sequoia, individual tree loss, when brought about by an abnormal accumulation of ground fuels, represents a critical threat.

Two important questions needing answers are: (1) what surface fire intensity will the fuels on the untreated area support under wildfire conditions, and (2) what impact on fuel reduction has been accomplished by prescribed burning? To answer

Table 2. Crown fuel characteristics before and after prescribed burning in a giant sequoia-mixed conifer forest.

Characteristic	Preburn	Postburn
Live crown weight—		
Tonnes per hectare	18.0	7.8
Tons per acre	7.2	3.1
Crown volume ratio (CVR)	108	142
Mean height crown base (MHCB)—		
Meters	0.9	4.9
Feet	3.0	16.0

Table 3. Distribution of representative packing ratios in a sequoia forest before and after prescribed burning.

Height above ground		Packing ratio (B)	
Meters	Feet	Unburned	Burned
1.5	5	0.000177	0.000028
3.0	10	.000289	.000023
4.6	15	.000204	.000049
6.1	20	.000276	.000107
7.6	25	.000249	.000127
9.1	30	.000199	.000128
10.7	35	.000198	.000091
12.2	40	.000133	.000079
13.7	45	.000122	.000083
15.2	50	.000092	.000070

Table 4. Fire spread model results before and after prescribed burning in a giant sequoia-mixed conifer forest.

Item	Preburn	Postburn
Forward rate of spread—		
cm/sec	3.8	0.05
ft/min	7.5	0.1
Reaction intensity—		
cal/cm²/sec	28.8	0.25
BTU/ft²/min	6367	55

these questions, we utilized the fire spread model developed by Rothermel (1972) . Rothermel presented eleven different fuel models which may be used to represent most fuel types. The appropriate model for the sequoia forest is timber with litter and understory. The preburn reaction intensity of 28.8 cal/cm /sec (6367 BTU/ ft /min) derived from our data is somewhat higher than the 18.l-cal/cm /sec (4000-BTU /ft / min) reaction intensity derived for this fuel model under similar ambient conditions by Rothermel. This could be due both to fuel type and accumulation under recent fire suppression policies of the area. It indicates potential for high intensity surface fires in the preburn sequoia stands studied and, combined with the low preburn height of the base of crown fuels, a consequent high potential for crowning out or torching of individual trees—including giant sequoia.

Using preburn and postburn ground fuel measurements (Table 1) and relatively extreme conditions for wildfires of 4 percent fine fuel moisture and a wind speed of 2.2 m/sec (5 mph), the Rothermel fire model predicted a substantial decrease in the forward rate of spread in ground fuels and reaction intensity following the prescribed burn (Table 4). These predictions represent essentially no forward spread under extreme conditions. Thus fuel reduction accomplished by

the prescribed burning greatly reduced the potential for high intensity surface fires that could lead to crown fires-even under relatively extreme ambient conditions.

The reduction of crown fire potential in these stands by prescribed burning has both short-range and long-range implications. Short-range reduction is accomplished through combustion of surface fuels. However, these fuels will promptly re-accumulate and will soon approach their former volume, perhaps in 5 to 8 years. Furthermore, the fuels from the standing trees killed by the fire will fall to the ground in 2 to 5 years and will accelerate the surface fuel accumulation process. This calls for multiple prescribed burns to reduce crown fire potential.

Long-range impact of the prescribed burning will be the reduction of live crown fuels at lower levels within the stand. The fires killed many small trees and lower crowns of larger trees (Fig. 1). The encroachment of shade tolerant species will continue in the absence of fire, of course. However, the fire removed much advance reproduction of shade tolerant species, and it may take 25 to 50 years for the fuels to develop to their former condition.

Conclusions

High intensity surface fires with associated "crowning" or "torching" of individual trees and clumps of reproduction threaten giant sequoia overstory. Although it has dramatic impact, a single prescribed burn is not sufficient to reduce this threat over a long time period. After long fire exclusion, several repeated burns, perhaps every 5 to 8 years, will be required to sufficiently reduce the existing volume of crown material in the understory, followed by periodic burns at longer intervals (probably 8 to 20 years) to prevent the development of dense growth of shade tolerant white fir so characteristic of a sequoia forest without fire. Ideally, initial efforts would be followed by prescribed burns or natural fires at intervals duplicating natural fire frequency.

Literature Cited

Biswell, H.H., R.P. Gibbens, and H. Buchanan. 1968. Fuel conditions and fire hazard reduction costs in a giant sequoia forest. Calif Agric 22(2):2–4.

Countryman, C.M., and C. Philpot. 1970. Physical characteristics of chamise as a wild land fuel. USDA Forest Serv Res Pap PSW-66, 16 p. Pacific Southwest Forest Range Exp Stn, Berkeley, Calif.

Davis, K.P. 1959. Forest Fire. Control and use. 584 p. McGraw Hill, New York.

Fahnestock, G.R. 1960. Logging slash flammability. USDA Forest Serv Res Pap INT-58, 67 p. Intermountain Forest Range Exp Stn, Ogden, Utah.

Kilgore, B.M. 1973a. Impact of prescribed burning on a sequoia-mixed conifer forest Proc Annual Tall Timbers Fire Ecol Cont 12:345–375.

———. 1973b. The ecological role of fire in Sierran conifer forests: its application to national park management. J Quaternary Res 3 (3) :496–513.

Rothermel, R.C. 1972. A mathematical model for predicting fire spread in wildland fuels. USDA Forest Serv Res Pap INT-115, 40 p. Intermountain Forest Range Exp Stn, Odgen, Utah.

Sando, R.W., and C.H. Wick. 1972. A method of evaluating crown fuels in forest stands. USDA Forest Serv Res Pap NC-84, 10 p. North Central Forest Exp Stn, St. Paul, Minn.

Storey, T.G., W.L. Fons, and F. M. Sauer. 1955. Crown characteristics of several coniferous tree species. USDA Forest Serv Inter mountain Tech Rep AFSWP-416, 93 p.

Van Wagner, C.E. 1968a. Fire behavior mechanism in a red pine plantation: field and laboratory evidence. Can Dept Forest Puhl 1229, 30 p.

———. 1968b. The line intersect method in forest fuel sampling. Forest Sci 14(1):20–26.

Wilson, C.C., and J.D. Dell. 1971. The fuels buildup in American forests: a plan of search. J For 69(8):471–475.

Publisher's Note

In some ways, this anthology was over 100 years in the making. The oldest article selected, "The New York Forest Fire Law" by C. R. Pettis, was first published in 1903. The newest, North et al.'s "Using Fire to Increase the Scale, Benefits, and Future Maintenance of Fuels Treatments," was first published in 2012. Without a doubt, the layout and styles used in SAF's journals over the span of several decades has shifted and changed gradually over time. Notice how all articles published more recently include abstracts, while older articles seemed to have adopted a more laissez-faire approach. Although the original articles are worth a visit if only to appreciate the differences in style, composition, and even copy-editing conventions, all articles from the archive selected for this anthology have been reflowed and composed from their originals to achieve a uniform look. Although the visual element has changed for this title, the text itself is presented here as it was published in its original form, except for in a few instances where the technological and human elements of this rather tedious recomposition process may have introduced minor typographical inconsistencies. We ask readers to keep these considerations in mind when reading through the archival content.